Merchant Navy: Letters from a Radio Officer

Ian M. Malcolm

FOREWARD	1
LETTERS 1951-53	2
LETTERS 1954-55	38
LETTERS 1956-58	62
LETTERS 1959-61	86
LETTERS 1962-68	113
LETTERS 1990-92	142
MOLLAND FAMILY HISTORY	147

Published in 2017 by
Moira Brown
Broughty Ferry
Dundee. DD5 2HZ
www.publishkindlebooks4u.co.uk

Letters from a Radio Officer was first published as a kindle book in 2013.

Copyright © Ian M. Malcolm

The right of Ian M. Malcolm to be identified as the Author of this work has been asserted in accordance with the Copyrights, Designs and Patents Act 1988.

All rights reserved. No part of this book may be reprinted or reproduced or utilised in any form or by any electronic, mechanical or other means, now known or hereafter invented, including photocopying or recording, or any information storage or retrieval system, without the permission in writing from the Publisher.

ISBN13: 978 1 52079 554 6

FOREWARD

The following are extracts from letters sent to me by Radio Officer Richard Edwin (Dick) Molland with whom I made two voyages to the Far East on Alfred Holt and Company's *Glengarry* during 1949/1950. On gaining his 1st Class PMG (Postmaster General's Certificate in Radiotelegraphy) at the British School of Telegraphy Ltd., 179 Clapham Road, London in 1950, Dick did one voyage on the *Glenearn* before resigning from Holts. Although, by law, it was necessary to carry only one radio officer on their ships, Holts preferred to carry two, with the 1st R/O as purser. This was why Dick resigned. Although an extremely competent radio officer, he felt, and I believe rightly so, that he could not cope with the purser's duties.

From 1951 until the 1960s, Dick worked for Brocklebank, Marconi, Redifon (ashore and afloat), the Crown Agents, Clan Line, the RFA (Royal Fleet Auxiliary), Ferranti (in Edinburgh), and Marconi again before settling for a shore job in London.

Although Dick never found a job which suited him, he certainly saw a great deal of the world as he visited India, the Persian Gulf, Egypt, South and East Africa, Canada, the United States, Hong Kong, Singapore, Australia, Venezuela, Trinidad, Bermuda, Italy, Lebanon and Continental Ports. And few can claim that they have dined with the Duke of Edinburgh in Antarctica.

LETTERS 1951-53

Dick has just joined Brocklebanks after leaving Holts.

Chelsea, 11th October, 1951

The coasting in the *Manipur* lasted about a fortnight. From London we went to Southampton, then Avonmouth, Middlesbrough, Antwerp, and back to London. I stayed on to help with installing some new gear and lived at home. I have been free the last couple of days and am awaiting further instructions.

The *Manipur* is a single screw turbine wartime effort which does 15 knots, and being a wartime effort it is hard to compare with Alfie's (Alfred Holt & Co.) newer ships. The *Telemachus* is a wartime effort, and I understand is pretty grim inside. To get back to the *Manipur* – there is running cold water in all the rooms, but not hot; the rooms are all very spacious. Apart from one other Brock ship (a post war one) which I have seen, the wireless room is quite the finest and most comfortable I have ever seen or worked in; it puts even Alfie's newest ships to shame from the point of view of comfort and operating convenience. The gear was a bit old: Siemens M.F. transmitter, Siemens master oscillator controlled H.F. transmitter, two of Brock's own design receivers – one for L.F. and M.F. and a separate one for H.F. But that has been altered now: we put in a Marconi Reliance (mains operated) which is the main M.F. transmitter, emergency set. The Siemens H.F. transmitter remains for the time being. The Direction Finder (579) is in the wireless room (wireless room is on the bridge) and there is a gyro repeater near the goniometer. There is a port at either side of the room. The desk is colossal, there is ample storage space, a little work bench, a chair, a two-seater bench (with a back to it) and a full length settee. There is a "serving hatch" into the chartroom.

There is a loudspeaker intercomm. system to the wheelhouse, radar hut on the monkey island, an ordinary telephone to the No.1's room, a call bell to the No.2's room and call buzzer to the wheelhouse. There is an inside stairway so there is no need to go outside. There is a spacious battery room on the boat deck complete with running cold water in a lead-lined sink. Two decks below the wireless room near the pantry is the motor room where all except the emergency motor alternators are installed. There are the following machines: radar machine, gyro compass machine, two big motor alternators for transmitters, and two small rotary convertors for the A.C. (230 volt 50 cycles) supply to the wireless room for receivers, frequency meters, and so on. Emergency machines are installed in the wireless room. All radio machines are of course controlled from the wireless room.

For some years before the war Brocks have been installing all machines (except emergency) one or two decks below the wireless: this keeps induction interference down to an absolute minimum as well as mechanical noise. Maintenance is also much easier as you don't have to be a contortionist to get into an inaccessible motor cupboard. Brocks are definitely radio conscious, and we are treated more as technicians than operators. They have an old coal burning "Fort" boat (the S.S. *Makalla*) which has even been fitted with Oceanspan. Brocks were the first company to have their whole fleet (including old wrecks) fitted with radar.

It is a change to see the mate and the skipper in their respective rooms slogging away at typewriters while we can walk around smoking pencils if we feel like it! (*Neither Dick nor I smoked, but, when in a jovial mood on the Glengarry, we pretended to smoke pencils.*)

Chelsea, 12th April, 1952

I don't know whether I told you in my letter from Calcutta, but on our way out somewhere near the Portuguese coast the *Glengarry* roared by quite near on her way home for Christmas: I could have wept! We had rather an unsavoury Christmas in Aden: the Old Man and the Mate started having a row in the morning which was carried on and brought to a climax in the saloon at lunchtime, and ended up with the Mate walking out in disgust and one of the Scots engineers also walking out in high dudgeon. It was all very embarrassing and distasteful. Needless to say most of the bods were fairly well oiled by the time Christmas dinner came round, and the Old Man did not appear. It was not what one would have called a happy voyage – that applied to everyone, even to the drunks. The whole atmosphere was very different even from a Blue Funnel [1]"Sam" boat, and not an atmosphere in which I could feel happy or at home. I don't think anyone can call me narrow minded, but there are reasonable limits to everything.

Of course, there is the other side of the matter: the run is so awful, and the ships so uninspiring, that if anyone is stuck there for good, I really can hardly blame people for more or less going haywire and really hitting the bottle excessively as that is about the only escape. The general standard of personnel in Alfie's is very much higher, but then I suppose if Brocks tried to tighten things up at all, they just would not be able to man their ships even though they have only got 23 of them. They seem to have great difficulty in getting engineers, and most of the junior mates are just getting sea time for

---

[1] *These were Liberty Ships leant to Britain by US under a Lend-Lease Agreement during the war. Their names all bore the prefix 'Sam', and many were subsequently bought by the companies which managed them during the war.*

master's tickets prior to taking up pilotage, and so on. My No.1 told me the Radio Superintendent told him not long ago that he could place 12 operators immediately! So there you are.

The food was absolutely chronic, and I had more or less permanent indigestion from the time I left home till the time I got back again. In Calcutta things were particularly bad because the crew is paid off there, and the cooking is done by shore gash hands. There is no supervision in the galley as there is no 2nd steward, but only a chief steward cum purser. I went to the galley just before lunch one day to get some water in my can and was nearly sick when I saw the filthy water and rags which were being used for washing up the cooking utensils. On more than one occasion the Mate caught them washing up in dock water, and that dock water makes London dock water appear to be distilled water by comparison! I was told that on one ship soup was made with dock water with the result that there were two or three deaths amongst the Europeans on board. Indians can make curry and cook rice, but that is about all. Apart from anything else, you get cargo wallahs and coal wallahs who seem to be in the last stages of T.B., and they go about depositing the contents of their lungs all over the decks, and usually just outside your door. I told one of them off, and he seemed genuinely indignant! Then there is the question of mosquitoes.

Coaling used to go on day and night, one basket at a time, and every lump seemed to feel as if it hit my bunk while it bounced its way down to the bottom of the bunkers. All night long at sea it would feel as if coal were being shovelled out from under my bunk – the bunkers were under my room! The next trip is going to be in the hot season, and I really feel sorry for all who will be condemned to sail in that old wreck.

I wouldn't mind a little discomfort if I could see some future in things, but there certainly was no future there. It would take me 5 years to reach the pay I was getting with Alfie after 3 years, and the promotion policy is such that one would be a No.2 for a minimum of 3 years. Even then, they have now got to the stage where it is just a question of waiting for dead men's shoes. Ships built in 1937 and 1939 have no running water in the rooms, but have only just had cold taps fitted in the mate's rooms (1st mate only of course) and a proper shower in the bathroom. The era of H & C started around 1947 or '48.

It seems to me, though I suppose it would never be admitted, that Brocks earn the money carrying cargo, and that money goes straight into Cunard to pay higher wages, bonuses, better ships, etc. You see, the chairman of Brocks is the brother of the chairman of Cunard, while John Brocklebank himself is only a mere member of the Brock Board.

Anyway, I did do something which I would not have been able to do in a wireless company, or any other shipping company for that matter. We had a Siemens H.F. transmitter – a master oscillator job which the majority of Brock ships old and new have. It makes rather a shocking noise on the air and you have to juggle with several controls at the same time. To cut a long story short, I re-hashed the output stage and aerial circuits converting it from a swinging link series tuned effort to a parallel fed anode circuit with pi-coupling output network as per Oceanspan and old established (*radio*) amateur practice. Tuning was simplified and the results were excellent as far as range was concerned. Other Brock ships up to 4,000 miles away (16 mcs) whom we could hardly hear were giving us good signal strength reports. I also managed to improve the note by putting in additional smoothing which consisted of the smoothing unit from our emergency Marconi 533 transmitter. I wrote up a screed complete with circuits which my gaffer copied out and typed out and sent in to the Liverpool Office together with a summary of contacts made with other ships and shore stations. So whatever they may think about me, they can't say that I took no interest in my work!

Our third mate knew the third mate of the *Phemius* (ex *Alcinous*, built 1926) and was astounded to find running water in the rooms! You see, although it may be 'industrial snobbishness' on our part, the majority of Brock blokes just have never been on board a decent ship in their lives. The percentage of decent ships coming into Calcutta is relatively small, just the odd new Clan boat or City boat, both of which are beaten hollow by the Glens. About the nicest ship going there now is the *Eastern Queen* which is a lovely brand new Jardine ship. It is not unlike a scaled down and slightly modernised Glen boat. So amongst that lot, the Brock ships seem quite good.

When I was coasting on the *Manipur* – one of the very much better ships – someone asked me what I thought of it. By the manner of asking it was obvious that my answer was expected to be very good. But when after a little thought I said that it wasn't too bad for a wartime utility effort and that it did not compare with Alfie's ships, the bloke was most surprised.

We docked at Tilbury on the 26th March, and I came home the same day. Within a couple of days I had sent in my resignation to Liverpool. I had already asked my gaffer to intimate to them in an official letter that I might be leaving soon as I found the run did not agree with me. Which of course was perfectly true.

Apart from the satisfaction of having got a 1st Class ticket, the thing seems to have been rather a dead loss, with the result that what satisfaction I had is beginning to wear thin, to say the least! You remember I tried N.Z.S., and that came to a sad end because of my eyes? Well, I tried Union Castle this time, and that has also come to nought. Owing to their policy of promotions by strict rotation, they will not take anyone older than 23. That only leaves the wireless companies and Smiths of Cardiff and Bibby. Smiths I don't fancy somehow, and the Bibby run to Rangoon is rather a dead loss, and they still run some pretty antique ships. I wrote to Marconi's and gave my reason for leaving Brocks as the run to Calcutta with the long stay there being unsuitable. Well, I got a reply by return post enclosing an application form but pointing out that their people are required to serve in any part of the world. You see, I would very much like to try a trip or two on one of the big P. & O. or Orient ships, and of course the only way of doing it is via GTZM. (*Meaning the Marconi Co. as this was the collective call sign they used to send a message to all their employees on ships at sea.*) But I am leaving that in abeyance for a day or two while I am investigating the possibilities of a job with Redifon here in London. They make marvellous marine and shore gear. Of course, I can only expect a very junior sort of job to start with, even though I have known their marine transmitter designer since 1942 and worked with him in the same laboratory for some time during the war.

To get back to Marconi. They are in a semi-permanent state of having between 200 and 300 vacancies. Now, with a 1st Class ticket and reasonable sea experience, and by not appearing to be very particular whether they give me a job or not, it might be possible to flannel my way onto a really good ship. But eventually there will be the inevitable attempt of the Bombay [2] transfer to some B.I. (*British India Steam Navigation Company*) wreck for a 2 year session, and that means resigning from yet another job.

I wrote to Alfie telling them that I had resigned from Brocks, and said that if in the foreseeable future they should decide to

---

[2] *It was a practice of the Marconi Company to place a junior R/O on a passenger liner bound for India, and, as soon as she arrived, transfer him to a vessel based on an Indian port. A 2-year stretch was then the norm.*

split the jobs of purser and radio I would be glad to be considered for re-employment. I didn't get the usual return of post reply that one usually gets from Alfie, and I thought they would just ignore it, and would not have been surprised or offended at that! But eventually I got a letter from WAC (*the initials of Mr Calverley who dealt with radio officers*) regretting that they cannot reinstate me in a purely radio capacity! It seems a little ironic that if you have a radio ticket you can't have a radio job, doesn't it?

On the other hand, if I had stayed on with Alfie, the time would surely come when they would say that I must go No.1 or else. And in all seriousness I know that I just could not cope with the accounts, and would definitely go round the bend. I am also fully aware of my limitations when cargo work crops up as a No.2, and that at times I got to the stage when my attitude could hardly be called co-operative, and I do appreciate how quickly these little episodes blew over and left no hard feeling! Still, looking back on my two trips in the *Garry*, I think I can say in all honesty that we coped with things pretty well on the whole and everyone's satisfaction.

I wonder what Calverley and Price would say if we sent in a joint letter asking to go back on the *Garry*? It would certainly cause a stir if nothing else. But I don't know that that would really offer any solution to the problem. The whole thing is that the other chap's job always seems better than your own!

As we were approaching Colombo on our way home I heard the new *Atreus* on the air. She was homeward bound on her maiden voyage. I listened in and discovered that I had met Lennon, the No.2. He joined Alfie at the same time as I did and we were both standing by the *Denbighshire* together in London. I called him up and we had a 20-minute chat. We were in Colombo together, and I went over and had lunch on board. The amusing thing was that I went in the agent's launch (same agents for Blue Funnel and Brocklebank) with our Old Man who was on his way to another Brock ship. The Old Man always used to say what fine ships Blue Funnel ships were whenever we saw one, which was quite often, but he always ended up by saying that Blue Funnel certainly have ships, but that he would not like to work for them as they were too strict whereas with Brocklebanks you could get away with murder. I used to feel like replying that if you know your job there is no need to get away with murder! I discovered afterwards that he was thinking more in the line of [3] 'cumshaw', and believe me, my eyes were opened! 'Cumshawing' is a major part of the business for masters and mates, and of course chief engineers can make fortunes on coal burners.

In spite of everything, Brocks certainly have a good name in the outside world, and I am very glad I have tried them and found out things at first hand; as an experience it has not been wasted, but it was not an experience to be prolonged. I can only say I am thankful that they turned me down in 1948 when I only had a 2nd Class ticket. It may be a negative way of looking at things, but there it is.

I am paid up to the 20th of this month after which I become one of the unemployed, so I really must start thinking hard. The Post Office seem to be short of people for the coast stations, but the pay is so poor that it is hardly surprising. The basic pay is something like £5-£10 a week, and the rest you make up on overtime, which is not very consoling especially if you have to pay for digs in Landsend or Wick.

I wonder what Duncan (*Captain Duncan MacTavish*) would say if he saw us both marching up the gangway of the *Glengarry* in London. I wonder if he has had any good typhoons recently? (*We had been in one.*) I will let you know what transpires and whether I end up on a "Baron" boat or not!

---

[3] *The word used for a bribe in the Far East.*

(*The Baron Line, owned by H. Hogarth & Co., had a bad reputation and was known as Hungry Hogarths.*)

Dick was now employed by Marconi. the main wireless company which hired R/Os and equipment to shipping companies.

S.S. *Lucky Star* (*T2 Esso tanker,
Panamanian registry, call sign HOEI*)
At Sea, 25th July, 1952
QTH (*Position*) noon 23.18N 61.29W
Bnd Hamburg from Venezuela!

Your letter was just timed right to meet me at Fawley (Esso Refinery near Southampton) at the end of my first trip to the Persian Gulf – Ras Tanura (HZA). (*Call sign of the radio station*).

There were quite a few crew members for this ship travelling from the U.K. The 3rd Mate and myself were the only officers and travelled 1st class at Esso's expense all the way from London. The rest of the crowd went 2nd class. We went from Liverpool Street Station to Harwich, then on a Dutch ship to the Hook of Holland, stayed the night on board (alongside at the Hook) and left on the Scandinavian Express next morning for Bremen. We arrived there in the afternoon and changed to a local train for Bremerhaven. A bus was waiting for us there to take us with all our gear down to the ship. She had just come out from annual dry-docking and overhaul. I expected at least two days respite before leaving, so that I could get things properly organised, but no, we sailed the following afternoon.

My room was spick and span, but the wireless room was a shambles with the German shore radio people working there, and German radio surveyors running about. A brand new

Cossar Radar installation was in the process of being completed.

Out in the river the following afternoon they did a complete D/F (*Direction Finder*) calibration with tugs swinging us round. Compass checking was also done. Then we were off. I had barely found my way round the gear, leave alone the fancy accounting systems due to the foreign registry, when telegrams were being brought in for transmission. I had no idea there was so much traffic handled in tankers, but more of that later.

By the way, we just pass Aden each way as we have no need to stop there. Fuel is taken on at Ras Tanura, or any of the other oil places, and the water capacity is so tremendous that the Bremerhaven water lasted the whole 6 weeks round trip without any restrictions whatever! During the war these ships carried about 100 men all told, but the peacetime crew is only 44.

The Persian Gulf was certainly warm: sea temperature 91°, but we weren't there long enough to suffer. One of Strick's (*the Strick Line of London*) was going to be at anchor near Ras Tanura for 3 weeks before beginning to unload! That is where the suffering comes in. We arrived there on the morning of the 19th of June, and left the following morning. During that time we discharged ballast, loaded 15,000 tons of crude oil, took fuel oil, and off away home. Very nice, but of course it cuts both ways: we were less than 24 hours in the U.K. when we got back. All I managed was a 'phone call home – not even any shopping.

The entrance to the Persian Gulf is very impressive, and would put the fear of God into a Blue Funnel skipper: there are numerous islands through which you dodge and miss literally by tens of yards. The 'countryside' is like a continuation of Aden, but near Ras Tanura (which just north of Bahrain) and up there it is just desert.

A loaded tanker in any sort of rough weather is more like a submarine. The saloon is right aft and you risk a good soaking. The Arabian Sea was rather unpleasant with the monsoon, and we had to slow down. If she is pitching at all heavily, it feels terrible in the saloon, and takes some time to get used to. In the Arabian Sea on the way home I didn't go near the saloon for the best part of two days. Sandwiches were brought to me up here.

On the way home in the Red Sea I was listening to the *Clytoneus* working someone else, and discovered that one Leslie Bonenfant was No.1: he did his first trip as No 1 coasting with me on the old *Machaon* (ex-*Glenaffaric*) when we took her to the breakers. He then went to the *Demodocus* as No.1. A nice chap, very entertaining, but also very chaotic as far as office work is concerned. He had been 8 years with [4]I.M.R., and then with Pan American Airways. He did one trip as No.2 in the *Radnorshire*, and then come No.1 with me.

I called up the *Clytoneus* and sent a 'note' to Bonenfant. Then the No.2 says, "Molland ex GNCS? (*call sign of the Glengarry*) and I say yes. To which he replies, "QUA? (*Have you news of?*) Malcolm? He said he thought you had left Alfies'. It was O'Byrne who was with you on your Australian trip. I told him you were working ashore now, and he asked me to pass on his best wishes next time I write, so there you are. Seeing that I had never met O'Byrne myself, you sure must have told him some awful things about me for him to remember my name all this time!

---

[4] *International Marine Radio, a wireless company similar to the Marconi Co.*

We left Fawley in the afternoon of the 10th, again bound for Ras Tanura – or so everyone thought. Off the Portuguese coast we get a message saying we might be diverted to the Caribbean. It ended with "please maintain close wireless watch." A couple of days later it happened: "Now proceed Amuay Bay Venezuela." It was afternoon and Pop (*the Master*) was resting. I didn't even wait to type it out on a form, but took him my rough paper. Up he goes to the bridge, and we alter course by nearly 90° much to the amazement of a few bods out on deck.

No one had been to this place before, and we arrived after dark. I was hauled up for the Aldis lamp signalling, at the same time looking for buoys on the radar and optically!

The first night out from Fawley we missed cutting another ship in two by some 50 yards in fog! If this ship wasn't turbo-electric, and therefore capable of quick reversing, there would have been an awful mess. It was the other ship's fault – he whistled one thing, and then did exactly the opposite. About an hour or so after I went to bed the fog got pretty thick, and we were hooting. We could hear other ships hooting as well. I was tired, but couldn't sleep. Around 1 a.m. I heard this other ship hooting and rapidly getting very close. In fact I could almost sense his nearness: his whistle was just about as loud as ours. Both ships then gave 2 blasts, or whatever it was according to the rules to indicate which way we should each turn. He turned the wrong way, and I heard Pop shout, "full astern, hard a-starboard". I shot out of bed and put on my raincoat, more or less waiting for the bang. All I could see outside was churned up water. The 3rd Mate heard the commotion and looked out of his ports and nearly passed out: his room faces forward next deck down; all he could see was this ship right across our bows almost within jumping distance!

Around 3.30 a.m. I was called by the 2nd Mate for D/F. We were more or less lost as we had been stopped, we had been going astern, and we had been going slow. On top of all, the radar ceased to produce a picture, and there were lots of fishing boats around. The atmospherics were so bad that one couldn't really get an accurate zero. Anyway, I managed some reasonable fixes. An overload trip must have tripped on the radar after running for many hours, because when I switched it on out of curiosity it was quite O.K., much to my own amazement as well as everyone else. Pop heaved a sigh of relief, and we carried on. Then as dawn was breaking Ushant lighthouse flashed through a hole in the fog, and the Mate got a visual bearing of it. I went back to bed around 5 a.m. and dreamt of collisions and picking people out of the sea.

By the way, in Venezuela we loaded 15,000 tons of oil in just under 6 hours. Not bad, eh? One thing there is no noise or dirt from 'general cargo', and no damaged cargo, and no natives swarming over the ship!

The name ESSO is derived from the letters S and O which stands for S.O. which stands for the Standard Oil Co. of New Jersey: a tremendous concern with world wide connections. About 44 tankers are run by the British i.e. the Esso Transportation Co., Ltd. These include ships of British and Panamanian registry. Including all the subsidiaries in other countries (Dutch, French, Canadian, German, U.S.A., Italian, etc.) they own 250 tankers, including coastal tankers. Then there is the huge Esso refinery at Fawley.

The Panamanian registry is nothing to do with avoiding safety of life at sea rules or anything of that sort. We sign British articles and the Board of Trade does the paying off etc. These ships are recognised by the various seafaring unions. It is mainly a matter of business convenience. For instance, this ship used to be run by the Italian subsidiary: it was then transferred to the British without any need for re-registration etc. The British also run 3 or 4 big American built 'super tankers' – 17,000 plus tons gross, and 26,000 tons deadweight, and 16 knots service speed. There are 6 or 8 big ones on order in the U.K. You see, the British can run a ship for about half the cost per day as compared with a Yank-manned ship. Needless to say the Yankee Seamen's Union dislikes us intensely! Shell are doing the same thing: they have about 4 big Yankee 'super tankers', British manned and Panamanian registry. Yankee wages are so colossal that they can't run ships at a profit themselves. You can't have British crews running American registered ships, and the Americans can't very well present us with ships free of charge without domestic political upheavals, so the solution is Panamanian registry. Masters, mates and engineers do a 5-month tour of duty, and then get a month's leave, and then go to another ship. The last skipper we had was only 32! He is now on leave and we have a new one – a minute Welshman in his early 50s. We heard shocking reports about this bloke, but as far as I am concerned it is all hooey. He is not a man who will stand for any incompetence, and has put the fear of God into the catering department – a good thing too. But if you know your job, there is nothing to worry about. We are the best of pals, and he visits the wireless room several times a day for a chat. You see, Pop and I are the sole inhabitants of the midships boat deck.

<u>27th</u> I have a large room on the starboard side of the midships boat deck, forward of, and adjoining the wireless room with communicating door. My bunk is more like a shore-side bed than a bunk. As a standard wartime fitting there is an upper folding bunk, but that is no hardship as it is folded up against the bulkhead. I have a folding top writing desk, a chair for same, a nice easy chair, full length settee, and large wardrobe. I also have a large carpet with fancy design thereon. There are 5 lights: middle deck-head light of 100 watts, desk light,

shaving mirror light, and 2 bunk lights. Across the alleyway is my own shower and w.c. We all have our own bathrooms: the mates have theirs adjoining their rooms, the only thing is they have forward ports.

The 2nd Steward does the Old Man's place and my room, but an ordinary assistant steward does the mates' rooms. On the whole I should say that my status here is considerably better than either with Alfie or Brocklebanks. There is no suggestion whatever of my being an outsider and just part of the radio gear.

Incidentally, these ships have A.C. mains – 115v. 60 c.p.s. The wireless room has 440v 3 phase 60 cycles, 115V single phase 60 cycles, and 115V D.C. for battery charging. By using standard equipment the Yanks have wasted a good opportunity: with the A.C. supply, you don't need machines of course, just transformers and rectifiers. But as it is the machinery consists of a 1½ H.P. 440v 3 phase induction motor coupled to a 1250v D.C. generator at one end (H.T. supply) and a combined exciter and alternator at the other end. This alternator gives 500 cycles for modulation and filament supplies.

The M/F transmitter is crystal controlled: 807 oscillator, 807 buffer, and 2 large triodes (211s) in a neutralised P.A. circuit. Output is 200 watts. The H/F. transmitter gives complete coverage from 2 to 22 mcs – 807 oscillator (M.O. or crystal), 807 buffer, and 2 813s in parallel for the P.A. The output is 150 to 200 watts. The M/F receiver is a 4 valve straight set (long wave also) and the H/F receiver is a Scott.

This ship was built in 1943 before the Victory Ship 'single box installation' came out, so all the units are separate, but electrically similar to the other outfit. Apart from the H/F receiver, it is all R.C.A. gear. The A/A (*auto alarm*) is all electronic. The D/F has built in gyro-repeater and mechanical quadrantal error correction. There is a Rye V.H.F. R/T set in the chart room: it has a range of 30 miles and is for communication with the Esso office at Fawley. Tanker companies certainly appreciate radio. Due to the short stays in port, everything has to be done by radio: from ordering carbon brushes to new assistant stewards! I reckon I earn my keep her alright!

Being a 'foreign' ship, we can't use the [5]Area Scheme, but have to take the Portishead lists at the odd hours and then make direct contact for the traffic. I have worked Portishead on 22 mcs. (GKI) from off the south coast of Arabia in daylight; GKS (*16 mcs.*) from the Caribbean in daylight, and GKL (*8* mcs) and GKV (*6 mcs.*) at night from those parts.

---

[5] *The Area Scheme was a worldwide Fixed Naval Radio Network, available to ships registered in the UK and the Republic of Ireland, which operated between 1946 and 1972. Messages to ships were broadcast at fixed times and, as ships could send messages to any Area Receiving Station, there was no need to communicate directly with the one nearest the office of destination. (See letter of 18th April, 1957.)*

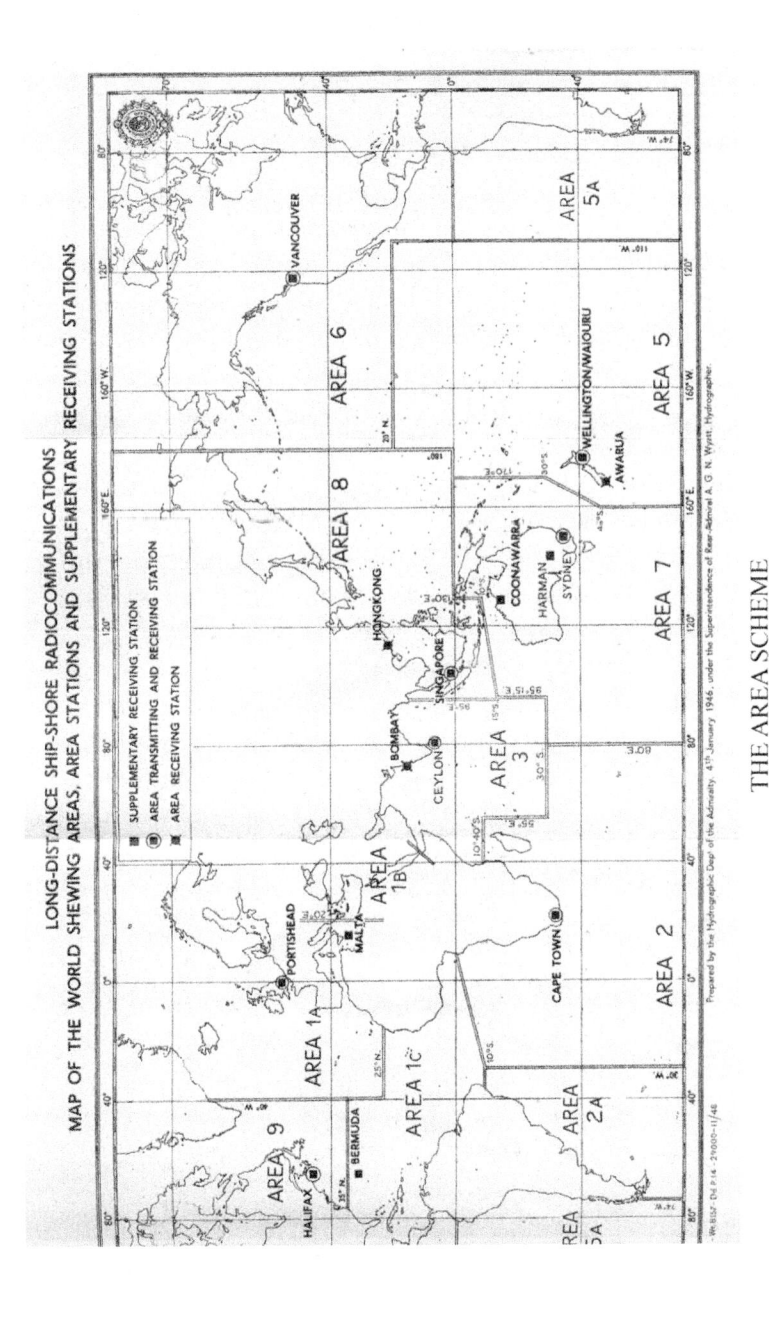

LONG-DISTANCE SHIP-SHORE RADIOCOMMUNICATIONS
MAP OF THE WORLD SHEWING AREAS, AREA STATIONS AND SUPPLEMENTARY RECEIVING STATIONS

THE AREA SCHEME

Accounting is a trifle complicated till you get used to it. As this is a 'foreign' ship, all transmitted messages have to be surcharged, but crew members pay what they would pay on a British registered ship, and Esso pay the extra. For these ships British C.C. is 6d, but L.L. is 1d and not 1½d. The share part of an SLT is 3/4d and not 4/2d, plus 5/- surcharge. I nearly went hairless the first couple of days trying to sort it all out. (*The charge for a telegram on board ship was composed of the ship station charge, the coast station charge and the charge for onward transmission by landline. Ship Letter Telegrams incurred a lower charge as the coast station forwarded them to the addressee by ordinary post.*)

The engine room is very effective with all the electrical gear. The propulsion motor is a 6,500 H.P. effort! 2,200 volts, 3 phase synchronous motor. A vast machine. By comparison the main turbo-alternator looks very small. The alternator is a 2-pole machine, and the propulsion motor has 80 poles, this giving an electrical gearing down ratio of 40:1. Speed is controlled by the speed of the main turbine, hence main alternator. The main turbine always runs the same direction: for reversing you reverse one phase. Hence the ability to do a very quick full astern. The Atlantic has been just like the Med ought to be, but seldom seems to be. We do up to 16 knots light ship, and between 14 and 15 loaded – according to weather.

Things are pretty free and easy as far as dress is concerned. You see, we can't send any laundry ashore, so we have to do everything on board ourselves. So next time you see a Panamanian tanker, don't think too unkindly of what may appear to be the 'scruffy load of Greeks on board'!

As far as we Marconi bods are concerned, it seems that one gets one's 'leave' once a year when the ship dry-docks for 3 weeks or so. However, we will attend to that later; it is a bit early to start thinking about leave. Anyway, I shall save plenty of money.

Pop said he can't understand why the larger shipowners don't all run their own radio departments. He was on one ship not long ago where the Marconi bod was a lad of 18 entirely on his own. If he really knew his job it would be O.K., but this bird was by all accounts a complete dope, and couldn't have ever seen a D/F set in his life before, leave alone used one. They were in fog in the Irish Sea: according to the D/F they were in the Bristol Channel one moment, and in the middle of Ireland the next moment – it wasn't even consistent. On top of all, he seems to have been a bit of an obnoxious bird, and Pop had him removed. That sort of thing unfortunately reflects on the rest of us.

29th - <u>Latest</u>: After Hamburg, Ras Tanura, Durban, Capetown!

Chelsea, 27th November, 1952

To be able to get dentistry done, I had to leave the *Lucky Star*. But there was a slight financial snag which I did not discover till after I came home. I was unavailable for duty and therefore placed on 'sick leave'. That meant that my pay stopped. Had I completed 6 months with [6]Guglielmos, they would have made up the difference between my National Insurance benefit and my full pay. But, as I had not completed 6 months' service, all I get is the benefit. As I am officially available for duty as from today, my pay restarts today, but, as they don't want me in Southampton yet, they have put me on a week's leave which will count against my annual leave entitlement. Had I been in a shore job, all I would have had to do is to take a few hours off at a time to go

---

[6] *Dick's way of referring to the Marconi Co. as Guglielmo was the inventor's first name.*

to the dentist and there would not have been any question of being off pay for two weeks. Or even if I was on a cargo or passenger ship which is in port for 2 or more weeks at a time, it could have been done without being off pay.

Being attached to the Southampton depot is quite a good thing, as they deal mainly with passenger ships, troopships and tankers. Although East Ham deal with all the big P. & O. and Orient Line stuff, they also deal with all the horrible things that come into the Thames. The shipping movements seem to point to my going on the *Empire Fowey*, unless it turns out to be a tanker. The *Empire Fowey* is due in Southampton around the 1st or 2nd December. She passed Malta homeward from Hong Kong on the 25th. The *Alcantara* is in Southampton now and the *Mauritania* is due in on the 29th from New York. Apart from tankers, that is about all. The usual thing with tankers is that the GTZM bod stays there till the ship goes for its annual dry-dock, and then he takes his annual leave! The bloke on board the Esso London has been there for 7 years!

Incidentally, Esso have about 8 jobs on order in British yards: about 17,000 tons gross with a service speed of 16 knots. The first is supposed to be ready for sea this month. One of these would be all right, but I would rather like to try a 'big ship'. At least there when you are off watch, you *are* off watch. Of course, I may be sadly disillusioned.

We had a very pleasant run down to Durban and Cape Town: it was a real pleasure to visit civilised places in a tanker and, what is more, we tied up at the normal commercial jetties. Of course, the snag was that we were only the usual bare 24 hours or so in each place. On top of that, we were all inoculated at Durban and I did not feel very bright as a result. It was spring down there at the time and Durban was quite hot in the middle of the day, but Cape Town was cooler and we wore 'home-side' clothing.

I thought I knew a fair amount about heat in the tropics, but I certainly learnt something new that trip. The Persian Gulf in August beats everything known and unknown and makes the Red Sea look very silly. The temperature was only 97 in the shade, but the humidity was colossal. All day long the steel decks were wet with dew except where the sun was actually shining. In the evening, what little breeze there was would drop and down would come the dew till the decks looked as if it had been raining: it was quite incredible. Just sitting very still, you would stream with sweat. It was hell, but it didn't last long, as we were soon away again. I certainly pity blokes on cargo ships that spend anything up to a month at a time in the Persian Gulf. Earlier on this year, there was a case of the mate of a cargo ship dying of heat exhaustion after he had been out on the fo'c'sle head mooring up for 4 hours in some out of the way dump.

The weather was a bit on the rough side after we left Cape Town on our way back to the Persian Gulf. In fact we were pitching so heavily we had to reduce speed. After turning the corner, we had a pretty heavy swell on the beam and rolled mightily till we reached the shelter of Madagascar. After that it was OK. From Cape Town we went to Mena el Ahmadi in Kuwait where we loaded over 16,000 tons of crude oil for the Fawley Refinery.

We stopped at Algiers for 4 hours to take on bunkers. On the face of it, that sounds silly, but in actual fact it is quite a good idea. Outward bound tankers take fuel oil to places like Ceuta and Algiers which is more sensible than being completely in ballast since freight is virtually free. That means that homeward bound tankers can take more crude oil and top up with fuel oil at one of these bunkering ports.

As usual, we were less than 24 hours in Fawley although we were 2 or 3 days at anchor in Cowes Roads as there were so many tankers waiting to go alongside the jetty. And all I managed was my usual 'phone call home. Anyway, the next and last trip was a short one: exactly 21 days.

It certainly was a long time from the end of May till the middle of November, but I think it has been worth it. Apart from anything else, I played with the radar both from the point of view of keeping it going or getting it going after it had stopped, and also from the navigational point of view. Being a tanker, it was virtually a continuous voyage from the moment I joined the ship. One good thing was the absence of cargo working noises and people on board working cargo. And it was great to be left completely in peace to do my own job! I think I can pretty truthfully say that I was happier than with Alfie's – no tin tallying lurking in the back of my mind. Certainly the money is less, but I remember thinking on the *Glenearn* that I would willingly stay on for £10 a month less if only there wasn't that perpetual cargo business.

One solution to the whole business would be for more of the bigger shipping companies to employ their own people thus breaking the virtual Marconi monopoly. Of course, the ROU (*Radio Officers' Union*) is to all intents and purposes a Marconi union, so where are you?

Chelsea, 9th January, 1953

My 'big ship' has not come off so far, but I suppose I mustn't grumble too much. Around the 8th of December, I got a telegram from the Southampton Depot about 5.30pm telling me to report at noon the next day ready for duty. Not knowing the ship I was going to, I had to pack all my belongings. I couldn't ring them up as the office was closed by then. I rang next morning before leaving home and was told it was the *Esso Cheyenne* which is permanently on coastal work and Coasting Articles. It was too late to repack then and there is always the chance of being transferred in Southampton. So off I went.

Although a 'coaster', it is a full size ship of just under 10000 tons and carries nearly 15000 tons of cargo – usually petrol! It is a wartime utility ship (British built in 1942) and was originally the *Empire Coleridge*. It is nothing to write home about after the *Lucky Star*. There is at least running hot and cold water in the rooms and there is plenty of heating and lighting. Of course, it is slow: only about 10 knots, but more usually 9. But that doesn't matter much when you only spend 24 of 36 hours at sea at a time. My room is on the bridge and the wireless room adjoins the chart room. It is a typical British wireless room with inadequate drawer space and motor starters arranged so that you hit your knees on them.

He then goes on to describe the layout and equipment of the wireless room which contained the Marconi Oceanic broadcast receiver, CR300 receiver, 398 H/F (High Frequency) and 381 M/F (Medium Frequency) transmitters, 579 D/F, and auto alarm.

We carry refined products from the Fawley Refinery to various distributing centres. At the moment we are carting around a lot of this new Esso Extra petrol which comes on the market next month. The first trip since I joined the ship was from Fawley up to Purfleet, just this side of Tilbury, and Saltend which is near Hull. I managed to come home for the day from Purfleet. The next trip was a full cargo to the Humber area and the one after that was to Avonmouth. Then we did two consecutive trips to the Thames, so I had last weekend at home and came home again yesterday evening. As we sail at 7am tomorrow, I shall go back to the ship tonight. The next trip may be to the Clyde or to Avonmouth, we don't know yet.

Most of Christmas Day was spent at anchor off Cowes and New Year's Day was also spent arriving at Fawley. If one gets home fairly often, this is quite a good thing, but otherwise I would rather go deep sea. The home coast is not very inspiring in winter! And being on coasting articles, we are on bare home-side rations so the food is nothing to write home about and the general standard of stewarding is rather low as no one worries very much about anything. Of course, there is no bond on board.

My predecessor was a mad Irishman who intended leaving Marconi's, with the result that the wireless room was in the usual administrative shambles. That is one thing that does annoy me. The Admiralty books have been uncorrected for so long that I just couldn't even hope to catch up with them, so I am leaving them. It was all I could do to correct the GTZM publications which my predecessor hadn't done. But I mustn't blame the man too much as apparently when he joined the ship it was a complete shambles. Whereas on the *Lucky Star* I had two filing cabinets and an office desk apart from the operating desk, all there is here is a very small operating desk and 4 drawers. Then of course the whole thing becomes a vicious circle: after joining several ships in succession which are in a shambles, one would no doubt feel why should one bother sorting things out nicely as long as the gear passes the surveys.

I have told the Southampton office unofficially that I would on the whole rather go deep sea, provided I get a decent ship. A coasting job would be ideal for a new bloke coming off a passenger ship and doing his first 'solo'. Sometimes these 18-year-old 'first trip solo' chaps get things round their necks when whisked away deep sea on a tanker, especially if it is Panamanian registry.

My pay is the large sum of £29.10/- per month! I left Alfie on £37. Brocklebanks would not allow anything for sea time and I dropped right down to £25, but an all-round increase brought that up to £27. GTZM pay B.O.T. (Board of Trade) rates and [7]sea time is counted. My actual sea time was under 3 years so I start on £29.10/-. I think that will soon go up to £31 once I have competed 3 years. It isn't much, I know, but then I started this game rather late in life. For a lad of 18 or 20 it would be quite good going. (Dick had been in the RAF during the Second World War.)

It has crossed my mind about going back to Alfie, but when I think it over carefully, and talk it over at home, I scrub round it. As it is I have left Alfie on the best of terms and I think it is best left like that until such time as old Lawrence Holt sees sense with the cargo worshipping aspect. It is no 'flannel' to say that my last trip with them on the *Glenearn* really got me down. I was hating every moment of every day and even my mother and grandmother remarked without any prompting that I did look somewhat worn and weary when I came home. Admittedly, the money was there and so was the leave, but is it worth going round the bend for the sake of it? In your case, you could cope with the job, and do it efficiently, but I know I couldn't.

I quite admit that GTZM was simply a last resort as far as I was concerned and, bearing that in mind, it has worked out fairly well. I can't alter my eyesight to suit N.Z.S. (New Zealand Shipping Co.) nor can I reduce my age to under 23 for Union Castle! Lamport & Holt have some shocking old things: I saw their *Lanlande* in Liverpool and I think I would rather have an old Brock ship! Of course, if you could fix it so that we both were to go back on the *Glengarry* or one of the P class ships, I am quite prepared to give the matter

---

[7] *Time spent on articles.*

serious consideration. When I see people waiting for buses in foul weather, waiting to go to horrible jobs, it has occurred to me that maybe I am better off even though I shall never get rich at this game.

S.S. *Esso Birmingham*
At Sea, 2nd April, 1953
ETA Port Said - 4$^{th.}$ Noon

GTZM at Southampton have been very decent. I was taken off the *Esso Cheyenne*, given 5 days leave and sent to this one. Actually, it was touch and go whether it would be this or the *Esso Belfast*: it was a question of the first one alongside at Fawley as both blokes were being relieved. Luckily, it was this one, as the *Belfast* is a pre-war Yankee job that does a full 9 knots! So I have a T2 (tanker) with a private bathroom and the No.3 of the *Alcantara* was being whipped off that ship for the *Belfast*. This ship is British registry (GTYS) which is more convenient in many ways than Panamanian. We seem to have very decent crowd here: the Old Man' is only 32.

This ship was built after the *Lucky Star* and is in some ways better and in others more utility finish. The gear is the big RCA single cabinet job with everything in it. The finish is sheet steel and it looks as if all the original crackle paint has peeled off and it has been repaired by hand. Anyway it seems to work OK. As you well know, Yankee stuff was never meant to last. Although the Marconi gear was crude electrically, it was certainly well made – in fact it lasts too long.

I believe I did tell you about my attempts to claim National Insurance money when I was officially placed on sick leave for dental purposes. The final outcome of that was that, under the Insurance Act, I was neither mentally nor physically incapable of work – and the fact that I could not work at sea and visit the dentist ashore at the same time just didn't cut any ice. I appealed to the Tribunal and was represented by the R.O.U., but still no go. But the final result was OK. GTZM agreed to pay me my pay less what I would have received from National Insurance, so I got £10 odd instead of £3 odd. I completed my 3 years sea time in January and now get £32.5/- a month.

You remember the famous East Coast floods? Well, I was on the *Esso Cheyenne* bound for the Humber from Fawley with the usual 14000 odd tons of petrol when we were caught in that storm. The Old Man took up residence on the bridge and was there for some 36 hours with meals brought up to him. We just had to heave-to and hope for the best. At one time we were drifting astern with engine going full ahead. The Spurn Light Vessel broke adrift and ended up miles away where they managed to anchor. A tug had brought a hulk from Singapore: it broke adrift off the East Coast, the tug's bosun was lost overboard and a man was seriously injured. I think the hulk sank in the end.

Just after 2am on the Sunday, the Auto Alarm went off and I was then in the wireless room for over 17 hours until we anchored in the Humber – just slipped out for a couple of quick meals. There was so much SOS traffic that it was impossible to keep track of what was going on. 1650 kcs (*the R/T calling and distress wave length*) as well as 500 kcs (*the W/T one*) were fully occupied. Humber Radio (*GKZ*) had been put out of action and (*Radio*) Amateurs in the Grimsby area were handling R/T distress on 1650 kcs on their own gear and were 'phoning up lifeboat and coastguard stations. (*R/T =Radio Telephony. W/T =Wireless Telegraphy ie. Morse.*)

The Auto Alarm was set off by a collier about half a mile to port – he was also sending SOS on his Aldis lamp to us. Apparently he was listing dangerously and was just about unmanageable. Actually, he got to Immingham safely in the end. We turned to try to get a little closer to him, but, when we got across the sea we did just about everything but turn over and were barely manageable ourselves. It would have been impossible to launch a boat in that sea and you would certainly never succeed in picking it up again. And with a full cargo of petrol, it would not do to have a collision. Good job nothing happened to us as the aerial downleads, main and emergency, were fouled up and I couldn't transmit. With green seas cascading onto us, there wasn't much we could do during the night. The Mate and I sorted things out next morning on the monkey island.

Contrary to popular belief, SOS was not the first signal sent by a ship in distress. The first was the Auto Alarm Signal which consisted of 12 4-second dashes with a 1-second interval between each dash. As the signal was made by the R/O on the Morse key, his timing could be out and, to allow for this, the Auto Alarm device, set to work on all ships when the R/O went off watch, responded to 4 correctly made dashes. A bell on the bulkhead above the R/O's bed, together with another one in the wheelhouse, would then be set off and, as the SOS message was sent 2 minutes after Auto Alarm Signal, he had just enough time to race to the wireless and switch on the receiver.

When daylight came, the wind started decreasing, but a heavy sea was running and it was the colour of liquid sand. We seemed more like the inhabitants of a matchstick than a full sized ship. I was all most alarming. Anyway, we got there in the end. The Esso head office in London were tickled pink that we had escaped damage. The Purfleet depot was flooded out: 5 to 7 feet of water. So no Purfleet 'voyages' for about a month. Exactly a week after that storm, we had a pretty good gale in the Irish Sea on our way to the Mersey.

Needless to say, I found the usual state of 'administrative' chaos here, although the gear and batteries seem well maintained. As usual, the Admiralty books haven't been touched. Volume 1 only came out this year and even that wasn't done. Volume 2 (D/F) is still the 1951 edition and not corrected at all. Volume 3 WX (*weather*) came out last year, but the oaf didn't order it till this year and, of course, no corrections. I had the new Volume 3 on the *Lucky Star* and left it up to date. Little or nothing had been done on the *Cheyenne* when I joined. I am hoping to do all Volume 1 here and Volume 3 as from my joining the ship. Volume 2 I shan't even attempt. I missed GYZ WX (*Gibraltar weather)* a few days for no other reason than the books being out of date.

I can't remember whether I told you before, but before joining GTZM, I went to the wireless school where I got my 1st and was offered the job of theory instructor. But I know I can't teach – especially electricity and magnetism from basic principles. The instructor who left was definitely a born teacher and very good at putting over mathematical problems. To try to step into his shoes and carry on where he left off would have been quite impossible.

The Cable & Wireless job was definitely no go as far as I was concerned. It means a life's work abroad on cable ships which are based in places like Aden and Singapore. You do 2 years and then come home on leave. Eventually you may get a shore job, but it would be abroad and to do with cables and not wireless. They only require a 2nd Class ticket so that you can do the ship's radio work, but your real work is on cables. Just fancy spending a few weeks at a stretch 'gardening' in the Red Sea!

I thought there were lots of opportunities for radio ashore, but my experience with [8]Redifon rather shook me. As far as the GPO coast stations work, that would be awful: you are really only a telegraphic machine, especially at Portishead. You never know where you will be stationed and you have to live in digs - and spend half your pay thereon. Give me a ship any day instead of digs.

About the Redifon receivers. I noticed that the *Glenorchy* had the rotary converter for the receiver in the wireless room which, of course, is a bad thing. To give Brocklebanks their due, even the old crate I was on had only the emergency machines in the wireless room. Alternators, etc., were all two decks down. The wireless room had sockets giving 230 volts AC supply. That is one drawback of this American gear: the transmitter machine is in the big cabinet with everything else and makes rather a lot of noise – mechanical noise. I must say these American D/F sets are handy things to use – you read the true bearing straight off a built-in gyro repeater scale and the Q.E. correction is done mechanically and automatically. *Quadrantal Error, introduced by in-phase re-radiation from the ship's hull, had to be allowed for as, otherwise, the bearing would be incorrect.*

I do agree that our job is worth more than we are paid, BUT that is provided the job is done properly and conscientiously. From what I can gather, that is rather the exception than the rule. I was sending a private telegram to another ship (not Esso, but Siemens I think). He was obviously making an awful mess of it by the way he was carrying on. Finally, he

---

[8] *Dick had applied for a shore job with Redifon and had been turned down. His assessment of Portishead Radio Station was incorrect. I started work there about 6 months after receiving this letter. The work was varied, you were never on the same job for more than two hours, and I was very happy there.*

apologised and said he had been on the beer the night before. Now fancy shouting that around on M/F in a busy part of the world!

S.S. *Esso Syracuse*
Bermuda, 15th October, 1953

We left the *Esso Birmingham* at Bremerhaven for her annual overhaul and dry-docking at the beginning of August and travelled home as passengers (1st Class all the way) from Bremen and via the Hook of Holland to Harwich route. I was supposed to have gone back to that ship and would have had about 3 weeks leave. But on the 13th day I got a telegram from East Ham at lunch time asking me to 'phone them urgently. I rang them at 2pm and was told that a rush (you ain't kidding) job had come up. The bloke who was supposed to come to this ship packed his hand in at the Shipping Office and wouldn't sign on and could I be at the Esso office by 5pm ready to depart by air for Palermo where this ship was in dock! It seems I was the only body within reach at the time. It will be at least a 9 month tour of duty. Of course, I could have refused on the grounds of short notice, but I didn't see that I should gain much by doing that. That could always get back at me by giving me more leave and then offering me the Indian Coast or a succession of dirty tramps. This way I put them in my debt and will have a good stack of leave due to me when I get home – and which I shall see that I get. Also, this ship is trading in American waters (more or less) which would be a nice change from the Persian Gulf in spite of hurricanes, one of which we met, and 'Winter, North Atlantic'.

Well, I did make it all right: nearly went hairless packing and arrived with all my gear at the Esso office in Regent Street, just before 5pm. The first half of the crew had gone out about two weeks earlier. A private bus took us to Blackbushe Airport, near Farnborough in Hampshire, and at 9pm we took off in a chartered 'Dakota'. We landed at Nice to refuel around 1.15am. We had to waste a couple of hours as you can't arrive at Palermo before daylight due to tricky approach through hills and lack of night flying facilities. We left Nice at 3.15am and arrived at Palermo at 6.15am. A private bus picked us up around 8am and we were on board by 9am.

This ship is a T3 and was built in 1942. Only about 20 were built (as against hundreds of T2s) and they were designed before the T2s although they are bigger and faster. This one is 11326 tons gross and has a horsepower of 8500 which gives us a fairly steady 15 knots. They were originally intended as motor ships, but the diesel engines weren't forthcoming in the States so they were redesigned to geared turbines. This necessitated some quick thinking as to the location of the boilers which has resulted in an unusual appearance for the after end of a tanker. The boiler room is at the front end of the after accommodation, on the same level as the saloon, and has doors opening straight out on deck. It is said that the firemen get more sunburnt on these ships than anyone else. The cargo capacity is between 17000 and 18000 tons. (*From 9.11.1942 until 21.1.1946 the Esso Syracuse was the USS Atascosa.*)

This ship had been specially modified for the New York-Aruba run. Two fridge cargo chambers have been built in the centre castle and an extra batch of fridge machinery fitted in the engine room. One of the cargo tanks has been completely isolated for the purpose of carrying drinking water to Aruba: separate pipelines and pumps were installed. Two cargo derricks have been fitted to the main mast and an extra winch or two. This is for carrying deck cargo to Aruba from New York. We don't have our own private showers and my room only has one port, which is a drawback. Anyway, the food is pretty good and very plentiful. And I have a comfortable bed.

The wireless room was in such a state of filth and disorder that I wrote to East Ham about it. The disorder was incredible. My motive in writing was not to get my predecessor into trouble, but to make quite sure that I don't get held responsible for other people's untidiness. Apart from anything else, it took me a full day till late at night just cleaning dirt and rubbish out of drawers and cupboards. You could hardly see some of the meter scales for grime on the glasses and white deck-head stand-off insulators were chocolate coloured with the accumulated grime of ages.

The people on board, including the captain, who had met my predecessor, were not surprised at the dirt and untidiness. Apparently the mates and engineers used to guess amongst themselves how many weeks the gentleman would go without changing his shirt! And yet when he went ashore he was dressed to kill. He was one of a large number of Southern Irishmen who work for GTZM.

From Palermo, we went to Sidon (Lebanon), which is a terminal of the Trans-Arabian Pipeline, and loaded a full cargo of Arabian crude for New York. The run across was very pleasant all the way except for the night of hurricane 'Dolly'. The worst only lasted about 4 hours, from midnight to 4am, but it was a very unpleasant 4 hours. I have never heard wind scream with such violence and force: it was as if something really evil had been unleashed upon us. The flying bridge became impassable for a while and it was impossible to relieve the wheel man.

That cargo of crude was actually discharged at Perth Amboy in New Jersey. It is on the far side of Staten Island. We then made our first run down to Aruba and loaded a full cargo of fuel oil for New York. We discharged it at the Metropolitan Petroleum Terminal which is right up the East River; under Brooklyn Bridge and several other bridges – and past the

UNO building. It was rather as if a full sized ship were to go up the Thames to Chelsea. Our second discharging place was the Erie Basin in Brooklyn. That is where we have come from now. We brought about 3000 tons of diesel oil for the local power station.

We are tied up alongside the jetty where the *Queen of Bermuda* ties up and it is alongside one of the main streets of Hamilton. A delightful change for a tanker. Even more remarkable: we came in this morning and will not be leaving till day after tomorrow. You cannot enter or leave in darkness. Owing to the smallness of the pipeline, it is going to take nearly 1½ times as long to discharge 3000 tons as it normally takes for 17000 tons.

This is a very unusual and attractive place: you come into a sort of lagoon with masses of little islands in it. Some of the islands just have one house with its garden and landing stage.

This is the longest break I have had on a tanker. I had a walk around this morning and went round the local broadcasting station, such as it is, this afternoon. Tomorrow I shall have another walk around.

Incidentally, this ship is Panamanian registry. There are 3 of the 17000 tons gross 'super-tankers' and 3 of these T3s out in this part of the world with British crews. Then there are several with Italian crews. A British crew costs 35% (official figures) of what an American crew costs. If you want to write direct, here is the address: S.S. *Esso Syracuse*, c/o Esso Shipping Co., Room 2300, Rockefeller Plaza, New York 20, N.Y., USA.

S.S. *Esso Syracuse*
New York, 15th Dec., 1953

I listen to the odd Area Broadcast, and make a point of listening to 1C at 1200 on Sundays for MBMSs. Being a 'foreign' ship, I take the foreign ship lists at the odd hour. I work Portishead quite regularly: every time we leave New York of Aruba, the 'turn round' report is send to London. As you no doubt know, we can't use the Area System, but have to make direct contact to send and receive traffic.

If Marconi's have any pleasant Indian Coast ideas in store for me, they are going to have to think again, because that is one thing I will not do. And what's more, I intend having a reasonably long leave when I get home, especially after "Winter, North Atlantic". It isn't that I am all that fond of tankers as such, but rather a question of playing safe (???). You see, there aren't many really bad tankers around these days, but there are plenty of shocking tramps with no H.F. which wander around the world on extended voyages. Marconi's can get all the people they want for passenger ships: old hands as No.1s and No.2s and first trippers as 3 and 4. It is the [9]H8 ships they have difficulty in manning.

---

[9] *H8 ships were those maintaining an 8 hour human watch with the Auto Alarm switched on at other times.*

LETTERS 1954-55

Chelsea, 16th July, 1954

We were supposed to have come home in the middle of May, but we had one or two diversions to Venezuela and a second trip to Bermuda which upset the schedule. The time before last in New York was when we more or less expected to be relieved. Although we had no telegram from London to that effect, the original written instructions to the old man had not been cancelled and he had everything worked out ready for signing off in New York – all to no purpose. Someone in the New York office must have slipped up that time. There were five Esso tankers (we were one) at anchor and almost every installation in the New York area had tankers alongside or waiting to go alongside. The result was that on the evening we arrived in New York, we were sent to Boston. We had lovely weather and made a slight change.

From Boston, we went to Bermuda and then to Caripito in Venezuela. That really is a place to end all places. You go into the Gulf of Paria between Trinidad and the Venezuelan mainland and then over 50 miles up the San Juan River. The jungle is not unlike the approaches to Port Swettenham except that it almost seems to be rather denser. There are quite a lot of crocodiles around that part. I saw four in less than five minutes.

From Caripito, we went back to New York. As each day passed without a telegram from London, we began to give up hope of being relieved at the end of that trip. Though the ship was due to dry dock next time in New York, I thought that we might hear when we arrived that we would take the ship to dry dock where we would be relieved. Then about 8 hours before we arrived in New York, we got a telegram from London saying that we would be relieved in New York and would fly home by chartered aircraft. Then the panic really started. The old man had to work out new pay offs and there was dense fog just to make things really interesting. I had my own little panic completing abstracts and damn silly things like statistical returns of traffic for [10]Macaroni. Still, this was the day we had been waiting for. We had a full day on board in New York before the new crew turned up and I was able to pack and complete my paper work at leisure.

On the afternoon of Thursday, June 17th, the new crew arrived. I showed my successor the works and left him in an electronic daze, but he should have got things sorted out reasonably quickly. He has a 1st Class ticket anyway. On top of all, I had typed out 5 foolscap sheets of 'technical instructions' and 2 pages of traffic notes. Besides, the ship was going to dry dock for about 3 weeks in New York as soon as it had finished discharging and gas-freeing. He had been with GTZM for some years, but that was his first tanker and he had never been with American gear before.

We left the ship around 6pm that evening and were taken into New York by private bus. Officers were taken to the Abbey Hotel and the 'boy's to a 'joint'. I was on the 18th floor. Standard Oil send all ships' officers to that hotel when they are either waiting to join ships or waiting to go home. It is quite a place: 8 dollars a night for room only, so I shudder to think what the Waldorf Astoria would cost!

The bus picked us up at 7am next day to take us to Idlewild (New York International Airport). Of course, there was the usual delay while the drunks were rounded up. The *Esso Richmond* (HOUZ) was relieved at the same time, so there were 64 coming home – 71 had been flown out. Most of the drunks seemed to be from the other ship. One AB was seen

---

[10] *'Macaroni' was a deliberate misspelling.*

lying in the gutter in Times Square at 3am with his head propped up against a fire hydrant. The *Richmond's* bosun was in such an alcoholic stupor that it was not till after we had landed at London Airport that he remembered his gear was still on his ship! We took off from [11]Idlewild between 10 and 11am on Friday 18th June and I was home for lunch on Saturday.

Esso had chartered a whole SABENA (Belgian Airlines) 'Skymaster'. It certainly was a vast aircraft – with 64 of us on board there were still several spare seats and all your gear went as well. We had a 4½ hour run to Gander where we stopped for about an hour for refuelling and light refreshments. Eats were also served on the 'plane. Then we had a 9 hour hop across the Atlantic to Brussels where we stopped for about an hour and then, finally, a 50 minute hop to London. Although we flew over London, we still had to go to Brussels. The reason seems to be that by international agreement airliners begin and end their official voyages at their home airports. If you fly to the States via KLM from London, you go to Amsterdam before starting out across the Atlantic. It all seems very foolish, but no doubt it has something to do with 'modern civilisation' – or something! All the flying was at between fifteen and sixteen thousand feet and it was pretty smooth all the way.

It was a lovely summer day in New York, at Gander and at Brussels. Soon after we crossed the coast on the last lap, everything disappeared in cloud and even the final approach for landing was made in cloud. We didn't see the runway till a matter of seconds before touching down. It had just been raining in London and it was anything but warm, but it was nice to be back. I was very glad to get away from the

---

[11] *Idlewild was renamed JOHN F. KENNEDY in 1963.*

American East Coast before the heat waves and hurricanes started.

The *Esso Richmond* had an even bigger panic as they were some 12 hours behind us and they were relieved at the anchorage. My opposite number from that ship told me that his successor was a lad of about 18 who 'knew it all' and couldn't to told anything. They were not due to dry dock till August and would be putting to sea again within about 24 hours, so I shudder to think what happened there.

My devoted employers very generously liquidated 38 Sundays at Sea and then gave me a month's leave which officially expires on the 20th. They still owe me 11 days after that. Last time I was home, they liquidated 45 or more Sundays at Sea. A nice state of affairs.[12]

I discovered that if one is on a ship which is permanently based abroad, one gets a 5% Foreign Service bonus. I asked the ROU man (Mr. D.B. Jones) at East Ham about this and he suggested that I ask at the Depot, but he warned me that 'they wouldn't know anything about it'. My next move should be to write to Chelmsford and, finally, to the Assistant General Secretary of the ROU. Well, he was right. The East Ham Depot didn't know anything about it as far as the *Esso Syracuse* was concerned, although two of the other Esso tankers on the American Coast were on Marconi's list of 'Foreign Service' ships (very significant in itself). So I wrote to Chelmsford. The reply consisted of one sentence stating that the *Esso Syracuse* was not considered as being based abroad!!! I think the so-and-sos would take the milk out of a blind man's tea if they could. And, of course, Esso

---

[12] *Home leave was generally given to compensate for Sundays spent at sea, but, where employers considered this impracticable, they could liquidate the entitlement by payment.*

themselves couldn't care less whether us 'Macaronists' get a 5% bonus or not. So I wrote to the ROU enclosing a copy of my letter, Chelmsford's and GTZM's reply to me. They wrote telling me that they had succeeded in having two of Esso's ships classed as foreign based and that they were taking the matter up with Marconi's[13].

I discovered that P & O started employing their own people some 4 or 5 months ago, so I wrote and [14]'spoke to them for a yob and they asked me no'. You must be under 30 and have a Radar Maintenance ticket at the time of applying. I think street sweeping will be next!

Chelsea, 3rd October, 1954

The very next day after the end of the official leave, I got the usual damn silly GTZM telegram telling me to report at 9am next day. It arrived just after 5pm and had the nerve to ask for acknowledgement. As usual no further information was given. I rang them up and got hold of someone in the office, but he conveniently did not know what was in store for me. I arrived at the Depot promptly at 9 and presented myself to the staff clerk - and the only intelligence I got out him was that I was to wait in the waiting room. About half an hour later, my name was buzzed out in Morse over the loudspeaker system and off I go to see the gentleman again. He still doesn't say anything sensible, but after a few minutes says, "You have a passport, haven't you?" To which I reply, "Yes" and had visions of going to Bremerhaven for another Esso tanker which would be on the U.K. run and which I would not have minded in the

---

[13] *Marconi's head office was in Chelmsford.*

[14] *This ungrammatical phrase referred to something we heard when at sea together. A Norwegian sailor, with limited English and relating how he asked for and been refused at job, apparently said, "I spoke the bosun for a yob and he asked me no."*

least. Then came the bombshell. "We're sending you to Karachi for the *Aronda*, says the gentleman! That started it. But I soon finished it. I asked him if that was going to be for a long session. He said it was, so I said, in that case, no can do. He slung a piece of paper at me to write a letter to the depot manager and state my reason. This was easy and took only about two lines. The reason was my mother's permanent state of health – she is far from dying, but she is far from well and I had no intention whatever of being away for 2 whole years. Mind you, I had no intention of going in any case. While on leave, I was joking that after 10 months on the American coast they will probably try to send me to the Indian coast. The run would have been Karachi, Colombo, Chittagong and then back again. That for two years is really asking rather a lot of even a semi-human being.

Actually, the *Aronda* (B.I.) is a nice ship - I have seen it in Colombo. It is one of B.I.s very few turbine ships in that part of the world and it is in the 18 knot class. It has radar and Oceanspan (a Marconi transmitter), etc. GTZM were even going to give me a private course on their radar here in London. Although only an 8000 tons ship, it is a Class I passenger ship and therefore must have a 1st Class ticket on board. That, of course, is really funny and serves Marconi right for their refusal to grant study leave. It has now backfired on them and they are short of 1st Class tickets. My opposite number from the *Esso Richmond* was there with me at the depot and he wouldn't have minded going at all, but couldn't as he only had a 2nd Class. So now they not only had to find another body, but a 1st Class ticket as well.

I was then told to go and sign on board the *Student* which they said was on coastal articles and was only going up to Middlesbrough and back. That is one of Hungry Harrison's Sam boats. And they certainly are hungry. There was another of their Sam boats in the same dock: neither seemed to have been painted since being taken over. They looked real filthy broken down tramps. All the accommodation is painted crew-accommodation green, including the old man's place. When I got on board, I found that the ship was signing deep sea articles, so I queried it with the old man and the shipping master and was assured that that did not matter as they could sign me off when we got back to London.

That was where I slipped up. When we got back to London, the purser (yes, they do carry one of sorts) said he had no authority to sign me off without authority from Harrison's London office. So I rang up GTZM and spoke to the staff clerk who was pleased as punch that I had signed deep-sea articles and had no intention of signing me off due to the expense of having me relieved, etc. So I told him that, as I had been tricked into signing on, he would have to find a relief in any case as I had no intention of going. This annoyed him and he suggested that I come up and see the depot manager. By the time I got to the depot the manager had retired home sick. The assistant manager was an oaf and I got no sense out of him apart from the bright saying that he had no authority to deal with the matter. I then told them that I wanted to take a month's leave off pay. But you can't even have that without a letter from a doctor or solicitor! So I got a doctor's certificate on behalf of my mother.

Soon after boarding yon tramp, I came to the decision that now was the time to get out of Marconi's. Up to then I had been relatively lucky, but there was obviously no future in the way they run the show. There was no question of starting on poor ships and graduating to the good ones. The situation changes hourly at the depot and it just depends when a ship happens to need a body. While I was down at East Ham on one of my visits during this palaver, two 'new entries' arrived. One, who had scraped through his 2nd Class, was going on the *Oronsay* on a Mediterranean cruise! No wonder no one gives a damn on GTZM ships and no wonder some of them are such shambles in the wireless room.

As I said, I intended packing up, but did not want to have an 'uncompleted voyage' in my Discharge Book. And if possible I wanted to leave GTZM on an amicable basis – officially at any rate. That part worked out OK.

A funny thing happened during my month's unpaid leave. A telegram arrived from East Ham asking me to ring them immediately, which I did. The staff clerk was overflowing with good fellowship and said that the *Esso Valparaiso* (British flag T2) was waiting for a body at Bremerhaven and would I like to go. I would have to travel the following night. I told him I would think it over. He stressed the fact that they were short handed. A couple of hours later, I rang him up and asked him no. In many ways I would have liked to have gone, but it would only be putting things off and making it that much harder for myself to get a shore job. A week before the end of my unpaid leave, I sent in my resignation.

Within a month after resigning from the Marconi Company, Dick became a Test Engineer with Redifon.

Chelsea, 9th January, 1955

When I have to wait any length of time for a bus on a cold wet day, I begin to wish I that I were some two or three thousand miles south, but when I get home in the evening I am thankful that I am not three thousand miles south or any other direction. What does get me down is getting up early in the winter: I have to get up at 6.45 a.m. and all the week I look forward to Saturday and Sunday when I don't have to get up at that unearthly hour. I used to get up regularly at 6.30 a.m. on the

Esso Syracuse, but that didn't seem so bad with the [15]'music room' adjoining my own room. However, I am not suggesting that I would rather be back on the *Esso Syracuse*! I find I don't bear any malice towards the clocking-in clock as being 'staff' we only clock in when we arrive in the morning. We only clock out if we work overtime, but the factory hands seem to do nothing but clock in and out: they clock out for lunch and clock in when they come back. But what really makes me feel like screaming at times is the public address system which in a factory like Redifon is a necessary evil as numerous telephone calls are coming in every few minutes for people who may be anywhere in the works, and that is the only way of finding them. The [16]P.A. system is operated by the telephone girls, and they seem to sound not unlike 'Miss Doris Plum' of 'Much Binding In The Marsh'[17].

My three months on probation have expired successfully, and the big noise told me they were quite satisfied with my efforts. Of course, being an industrial outfit, if the volume of work decreased, then anything might happen, especially seeing that it is not a vast industrial empire like Marconi or S.T.C. It certainly hasn't got the security or the pay of Alfie Holt, but that is neither here not there. As I said, there seems to be more than enough work coming in with plenty more pending.

As far as the premises are concerned, I am far from impressed. It used to be a gas mantle factory: I understand thorium is used in the process, and there is a danger of fire or explosion, and the result is that the main part of the works is in three separate blocks with outside stairways, gangways, and gantries linking up the three buildings. There are also a few large shed-like buildings. Our department, Heavy Test, is in

---

[15] *wireless room*

[16] *Public Address*

[17] *A popular radio programme*

one of these sheds. It would hardly do justice to a garage leave alone a radio outfit: the floor is concrete, the lighting is poor, and it is either too hot or too cold. The place is jammed up with gear, and there is nowhere to keep anything, and everything gets filthy. Hundreds of pounds worth of test gear get knocked around on open shelves. At times we overflow into an adjoining 'shed' which has a wooden floor and fluorescent lighting, but to make up for those innovations it is usually very cold. The canteen is a new building and central heating was completed just after the weather started getting really cold, but I am sorry to say that the food leaves a lot to be desired, and is rather expensive as well. As the puddings aren't too bad I take sandwiches with me for lunch which I eat in the canteen: I get a sweet and a choc-ice and coffee. Several other people do likewise.

Then of course there are the advantages: nobody stands over us supervising our every movement, and that in industry these days seems to be a rare pleasure. We had a reasonable break for Christmas: we closed at lunchtime on Christmas Eve, and didn't go back till the Wednesday after Christmas. So with a five-day week, that made a nice short week. Of course there are times when I wish I were back on the *Glengarry*. I must be honest and admit that I am somewhat lazy by nature: the very idea of having to work at all does not inspire me at all – yes I know I will never be rich, I came to that conclusion long ago! And I must say that I find a regular daily routine a bore and a bind, but then one can't have everything.

Chelsea, 13th October, 1955

I only came home on the 10th after doing a round trip to Canada on the M.V. *La Ensenada* (GTBN [18]QRC Redifon). It was awful, and I packed up after one trip, so I am again out of

---

[18] *Q Code for 'The accounts for charges of my station are settled by'.*

a job. Incidentally, I was QRY (*Your turn is*) 11 to Portishead on one occasion, and 6 on another, so I gathered things must be pretty hectic at times.

You may remember that I started with Redifon just over a year ago at £9 a week. Last April there were all round pay increases and my pay went up to £10, but in spite of that I was beginning to get fed up. The actual premises are awful: you freeze in winter and roast in summer. The work is quite responsible as we sign the 'electrical clearance' of all the equipment we deal with. At times it is very disheartening as the factory itself is on piecework and very often the inspectors pass stuff which they have no business to pass. That means we reject it when it comes to us, and much palaver and so on to get the job done decently. The test department is a wonderful place to start in provided you can later get into the development lab, but as a permanent set up it isn't much good. The absolute top rate is about £12.10/- per week.

By the time summer was approaching, the job was definitely beginning to get me down. Apart from anything else, one is on one's feet all day and by the time I got home in the evening I just used to have supper, read the paper and then go to bed.

Around June or thereabouts, I saw an advertisement in the Daily Telegraph for deck and radio officers for Swedish ships for temporary employment. It was a box number and I wrote in. I got a telegram back saying that if I was interested in one month's employment to report to an address in the City. I wrote and told them that this wasn't much use, but that if they could produce something a little more permanent I would be interested. Then they rang me up and gave me the details.

They wanted blokes for the Swedish Lloyd ships running to the U.K. and the Mediterranean, and the reason they wanted British deck and radio officers was that there was a dispute in

Sweden. Strikes are illegal in Sweden, so in an industrial dispute all hands just hand in their resignations – the net result being the same. That rather put me off: if British blokes were being taken on because of a real shortage of personnel, then we would probably be treated like lords. But if they wanted us only because they were saddled with an industrial dispute, then feelings on board might easily become rather strained. Although I understand that they are often genuinely short of seafarers of all kinds. They agreed with me that it would be silly to chuck up my job just for a month, but said they might have something for me in the near future.

A few days later the Swedish Lloyd agents rang me up at work: they wanted me for a new Swedish tanker on its maiden voyage. I would have to fly to Sweden to join it. It was going to the Persian Gulf in ballast and was doing three or four trips from there to Australia. They could guarantee me employment till October and the pay would have been not less than £60 a month with overtime on top of that. But there were several snags. First of all the dispute was still on. That was on Wednesday and I would have had to give a week's notice to Redifon not later than Friday. That meant I would still have to be at work all next week and yet I would have to go for medical examination, etc. They also emphasised the eyesight test. There just was not the time available to get everything done properly and I would have looked very silly if I failed the eye test. Having failed on that account as far as N.Z.S. was concerned, I had no intentions of running any more risks in that line. I told them I was sorry, but there just was not enough time. The bloke was very nice and asked if he might get in touch with me again should something else turn up in the future. Well, they settled their dispute and I heard nothing more from them.

In the middle of July we had our official two weeks' holiday and it was during the holiday that I saw an advertisement in the Daily Telegraph put in by Redifon. They wanted radio officers with deep-sea experience. This sounded interesting, so I went in and saw the bods in the Marine Department. As you may know, they have been running trawlers in competition with Marconi's for several years now. They are going to branch out into deep sea operating and a very rosy picture was painted of the prospects. The company they were dealing with was Buries Markes Ltd. whom I couldn't remember ever having heard of, but they assured me it was a very good company. They haven't got very many ships. They all have Spanish names: *La Ensenada*, *La Cordillera*, *La Pampa*, etc. The first ship they were getting was the *La Ensenada* which was built in Japan in 1950 for the parent company, Louis Dreyfus of Paris. This made me think a bit, but they assured me it was a lovely ship. I decided to take the job and was transferred to the Marine Department to become their No.1 deep-sea bod. The Chief of Test was not at all pleased so that I did not go and visit him when I was at the works the other day! Marconi's were operating *La Ensenada*, but very wisely refused to take on a maintenance contract for the Jap gear. Redifon agreed to do so and got the job. They are also going to refit and operate two old ships, *La Laguna* and *La Quinta*. I believe these two are French cast-offs from Louis Dreyfus. *La Quinta* has French gear at present, but is still being operated by Marconi.

The majority of Buries Markes' ships are new and are Marconi ships, so there isn't really much future for a Redifon R/O as far as they are concerned. Buries Markes' ships are all nice looking and beautifully painted and one gets a very good impression. In actual fact it is a shocking tramp outfit with atrocious feeding. One doesn't hear much of them because the ships don't often come to the U.K. They spend a long time tramping abroad and usually end up in a continental port. They specialise in bulk cargoes such as grain, coal, iron ore and sulphur. They don't seem to carry their own cargo, but are usually chartered out to other companies. For instance, *La Ensenada* is on charter to Runcimans of all people, but luckily she is on the Canadian newsprint run and is back in the U.K. every month. She is booked for that run till the end of next year.

Now to get back to the ship again. It was discharging at Deptford and we went down on Monday the 5th September to take over from the Marconi bloke. I was very impressed with what I saw. A nice looking ship of just under 6,000 tons with nearly 6,000 H.P. of engine. I am told the ship exceeded 20 knots on trials. At present they run it at a steady 15 in good weather; any greater speed would probably make it shake apart!

The ship was beautifully painted and the accommodation looked very good. My room was just under the bridge and next to the old man's; it had a square window facing forward and another one outboard. The wireless room was aft of my room and had a communicating door. All this was on the port side. The wireless room had two windows facing outboard and one facing aft. The gear was breathtaking. At the forward end of the W/T room was a combined MF and HF transmitter giving 750 watts on MF and 300 to 500 watts on HF according to frequency. At the other end of the room was an HF only transmitter which was very similar to the HF portion of the big one. You sat at a control table in the middle of the room which was not unlike a cinema organ console. There were three receivers across the front of the desk and remote control units etc., either side of you. Above the receivers were 15 meters which included remote reading aerial ammeters. Built into the desk was the emergency transmitter which worked on 8 mcs as well as MF. In the chartroom there was a 50 watt radiotelephone which also worked on the long distance R/T bands for which the ship is not now licensed. There was a rotating loop DF. The wheelhouse looked more like that of a warship with gyro-pilot and vast quantities of gadgetry. There was a Sperry radar. The main transmitters were remote controlled from the operating desk by push-buttons. At the bottom of each transmitter was a ¼ HP tuning motor. On HF you changed from 'call' to 'working' by remote control. The transmitters worked off 3 phase AC and there was a motor room full of all sorts of motor alternators with automatic starters. It really looked terrific. There was one big snag – it was Japanese. The materials were very poor and workmanship terrible. The HF receiver died on me on more than one occasion when I was working Portishead. Wafer switches kept jamming, or else didn't make contact and, one day, one of them just broke and I had to short contacts out with fuse wire in the teeth of an Atlantic gale. The French had made modifications to the keying circuits and had cut out the receiver aerial protective relays. The result was that the 750 watts of MF were backfiring into the MF receiver and had charred the RF stage coils! In spite of the large wireless room, there was an awful lot of space wasted and the desk space was quite inadequate. On top of all, the Marconi bloke who had been there some six months was a filthy, lazy, untidy devil – in fact an oaf of the first water. It took days just to clean up the place. He was so lazy that he even sharpened indelible pencils into the desk drawers! All this of course did not come to light when I first went on board.

Well, we left London for Quebec on Saturday the 10th September. The noise and vibration from the engine room had to be experienced to be believed. We were in ballast and were only drawing 11 feet forward and 18 aft. That alone is not calculated to inspire confidence for an Atlantic crossing at this time of the year. From leaving the Channel till almost in the Strait of Belle Isle, north of Newfoundland, we didn't have one single decent day. It was gale and storm all the way. One night I was almost hurled out of my bunk by the most incredible rolling. I could hear furniture and crockery crashing around. The steering gear had started to fall apart – we were out of control and stopped until they put it together again. A few days later we were in a terrific storm with north-westerly winds of almost hurricane force. I was really scared. We were more or less hove-to and were almost uncontrollable even with the rudder had over. She would drop away into a trough and then start rolling as I have never known a ship roll before. You could neither sit not stand, but just had to hang on to something. The old man's day room was an indescribable shambles and so was the wireless room. I really thought we were going over on several occasions. In the middle of the storm there was a momentary engine failure which didn't make me feel any happier. We eventually got to Quebec, where we spent two or three days, then up to Montreal and then down to a real one-horse town called Dalhousie, in a bay off the Gulf of St Lawrence.

We arrived back in Manchester last Saturday and I came home on Monday. In the past this ship had been signing six monthly articles and I went there prepared for a long stay, but luckily they had changed over to single voyage articles which, although not exceeding two years, had to be closed each time the ship returned to the U.K. I wrote from Quebec telling Redifon that I was not prepared to go back in that ship and, when I got back, I told them I was not prepared to go an any more Buries Markes' ships. So I had no alternative but to resign as the Marine Department is more or less a watertight outfit of its own. They managed to find a relief by the skin of their teeth and the ship is due to leave Manchester for Canada tomorrow.

As I said before, the food was atrocious, but added to that, the steam heating didn't work properly, the hot water wouldn't be hot for days on end, the wash water was dirty, the drinking water had lube oil in it and the sanitary water would fail for a day or two at a stretch! It was a feat to manage a hot shower.

After having had one cold shower I had to resort to stripping off in my room and washing in my basin.

Redifon are trying to get in on new Prince Line and Furness Withy ships, but I don't know when. They are going to have a hell of a job keeping blokes if they can only offer Buries Markes ships. They will just end up by getting all the down and outs and throw-outs from GTZM. It is a great pity because it could have worked out very nicely if they could have got in with a decent company. Anyway, I did see two icebergs, although we passed several which I missed.

They are only paying B.O.T. wages, but may pay a bonus of one month's pay at the end of the year. This would be called a maintenance bonus for looking after the gear satisfactorily. They started me off with one year's seniority as by the time we sailed I had completed one year with the firm. All sorts of rosy prospects were painted of the future, but I would not have gone back in that ship for £100 bonus for one more trip, and I really mean that. Incidentally, [19]AH are building more and more ships and they are being fitted with Redifon gear.

We even reduced speed less than half way home for no other reason than to economise on fuel – not because we didn't have enough to get us home, but to save money. That was one of the reasons why the heating wouldn't work properly. There was not enough exhaust gas to run the exhaust gas boiler properly and of course they wouldn't use the fuel burner in said boiler.

There was a very peculiar and indefinable unpleasant atmosphere on board though at no time did I have any disagreements with anyone there. It must be something

---

[19] *AH are the letters on the house flag of Alfred Holt & Co.'s Blue Funnel Line ships.*

peculiar to the Company. The mate was a peculiar bird: at times you would think he was staff captain of the *Queen Mary* the way he carried on. Anyway, it was a very interesting experience, but one not to be repeated. Just think if I had gone away on a two-year tramping voyage on that thing!

Incidentally, we brought back as passenger (supernumerary), from Montreal, the British Vice Consul in Paris who is a personal friend of the Louis Dreyfus family – he went to the States on one of the French ships. He told me that he intends telling them in Paris (the Dreyfus people) exactly what he thinks of the British set up. He said the food was marvellous on the French ship, but his remarks about the food on *La Ensenada* are quite unprintable!

I have written to the ROU telling them I have resigned and have asked them what the general situation is like nowadays with regard to freelance jobs. Resigning after one trip will not look good as far as Redifon's name is concerned.

Chelsea, 25th December, 1955

Had it not been for a technical hitch I would have been on my way to Antarctica last Thursday, but as it is we will not be sailing before next Thursday. The result is that I managed to get home for Christmas.

I am now Radio Officer/Purser of the new Royal Research Ship *Shackleton* of the Falkland Islands Dependencies Survey. QRC I.M.R., but am direct employed by the Government at £500 a year – same pay as the third mate. We do not pay U.K. income tax, but Falkland Islands' tax, which is some difference. Our call sign is GVDC at present, but when we get out there we will be re-registered at Port Stanley and we shall have a V call sign. The *John Biscoe* is VPNE.

I was still job hunting in November: I wrote to Niarchos, SAIT and even to the [20]Dockers. I had hoped to get one of Niarchos' new 'World' tankers, but there were none going at the time. They did ring me up to offer me a Sam boat at £75 a month, but I did not fancy that as there were already two of them in the Atlantic minus propellers. Carrying grain or coal across the North Atlantic in winter is not my idea of entertainment. Then one afternoon I was looking through the Official Appointments advertisements in the Daily Telegraph and came across one from the Crown Agents wanting an experienced R.O. for the new M.V. *Shackleton*. That was on a Wednesday. I wrote a letter that evening and posted it on the Thursday morning. At 5 p.m. on the Thursday the Crown Agents rang me up and asked me to go and see them on Friday at 11 a.m. which I did. I have never known a Government department move so fast. In the one day I was interviewed, filled in forms and was sent to Harley Street for medical and X-ray. On the following Tuesday I was actually on my way to Denmark to join the ship which was being more or less rebuilt at Frederikshavn in North Denmark.

The ship was built earlier this year by a Swedish yard for Norwegian owners and was called the *Arendal*. She was then of 950 tons gross. She had two hatches forward, each with its own electric crane, and the accommodation and bridge aft. The main engine is German M.A.N. of some 850 horsepower. The propeller is variable pitch and can be controlled from the bridge, as can the main engine. The ship was used for running to the Mediterranean and apparently carried 12 passengers as well. The speed was 11 knots which was not fast enough for the owners and they put her up for sale. The Crown Agents bought the ship and then spent £90,000 on rebuilding it. The whole superstructure was extended some 30 feet forward and

---

[20] *Ostentatious millionaires who owned a motor yacht.*

the second hold was done away with and incorporated into the accommodation. The gross tonnage is now 1102. All the accommodation except the chief engineer's room is completely new and very nice it is. In many ways it is more like a yacht than anything else.

We carry a total of 63 people: 31 crew and 32 supernumeraries. The latter are in two berth cabins complete with running H & C. The crew have better accommodation than on many a new merchant ship and of course the officers have single berth rooms. I expected my room to be combined with the wireless room, but it isn't. My room and the wireless room are up on the bridge.

Although we are all civilians, as it is a Government ship a certain amount of Naval practice seems to be in evidence. For instance we have a wardroom in which the engineers do not feed, although I think they are allowed to use it as a lounge provided they are properly dressed. The captain, three mates, ship's doctor and myself feed in the wardroom. There is a spare place for the Governor of the Falkland Islands when he comes to visit us, as he will do. The supernumeraries have a nice mess room of their own, and a very nice lounge of their own.

We also have special braid and cap badges, although the tailors could not get mine at the short notice they had so I will have to wait till next summer when we come home. The captain has three broad rings surmounted by an imperial crown, the mate 2½ rings with crown, second mate 2 rings with crown, third mate 1 ring with crown and I will have one ring on green background also with crown. The cap badge is something like a cross between R.F.A. and R.N. The ship is painted grey with a yellowish funnel. There is a crest on the bow with the letters F.I.D.S. on it. We also fly a special ensign. The crew are all provided with naval style caps and jerseys with 'Shackleton Port Stanley F.I.' across the fronts thereof. The bosun has naval style petty officer's uniform.

Marconi's couldn't produce the gear required at the relatively short notice, so I.M.R. got the job as they were able to produce the stuff in time. The wireless room is minute, but is full of stuff. I have three transmitters and three receivers. The main transmitter is the I.M.R. 81/83 which is their latest effort. I prefer the looks and finish of Marconi and Redifon gear to I.M.R. which is obviously built down to a price more so than the other companies' gear, but it is quite good and certainly very easy to use from a purely operating point of view. The main transmitter consists of separate MF and [21]HF units built into one cabinet. The [22]PA in each case is an 813 and the power output is 150 watts, dropping to 90 watts at 22 mcs. The main receiver is the huge I.M.R. 54 which is now fitted in the Queens and the new Cunarders like the *Saxonia*. It looks a very big brother to some of the Eddystone receivers. The reason is that Strattons developed and manufacture that set for I.M.R. Then there is the emergency transmitter and emergency receiver – also I.M.R. I also have the original Danish R/T outfit. The transmitter has four 807s in the PA, gives 100 watts output and can run off mains or battery. The receiver runs off batteries through a vibrator unit and is a very good general-purpose set which covers 100 kcs. to 22 mcs. It's also used for DF in conjunction with a rotating loop. Apart from the ordinary R/T frequencies we go up to over 6 mcs. which is used for working our bases in the Antarctic. In the chart room is an I.M.R. broadcast receiver which feeds five loudspeakers in various parts of the ship. A microphone is also supplied so that announcements can be made. The net result is that we have quite a collection of aerials strung up

---

[21] *High Frequency*
[22] *Power Amplifier*

which gives the impression of a young warship. Incidentally, the ship itself is only 180 long.

I travelled to Denmark via Harwich to Esbjerg at the beginning of the month and sailed for Frederikshavn on the 4th for Southampton. We had an awful trip; down to 3 and 4 knots at times. It was gales all the way and we were only drawing 5 feet forward. I was completely paralytic with seasickness: I never knew it was physically possible to be so seasick! Everyone except the captain, who has been on this stunt for quite a number of years, felt bad to a greater or lesser degree. Our second mate is an ex-Blue Funnel third mate who got his mate's ticket and left AH specially to come to this job.

On Monday the 19th there was a big do at Southampton when the wife of the Governor of the Falkland Islands performed the renaming ceremony, after which the captain introduced us to the good lady. Then there was a buffet lunch on board for the 60 plus visitors which was organised by a catering firm. The BBC people filmed some of the proceedings, it was on television that night and the captain spoke on Radio Newsreel at 7 p.m. Of course the local paper in Southampton has been full of us. We were supposed to have sailed on the 21st and the BBC were ringing up to ask when we would be leaving as they were sending a television film unit down, but we couldn't sail as we had to take on pig iron ballast.

When we do leave we shall be stopping at the Cape Verde Islands for water, then on to Montevideo and finally to Port Stanley. We shall be there for about a week and then we start visiting all the bases to land stores and relief bodies. We shall be right in the ice in spite of the fact that it will be summer down there. In their winter the bases are completely isolated as far as any shipping is concerned. The ship is based on Southampton and we return there next June. We will be there till October when we leave for the South again and then return the following June. During our stay in the U.K. we take our paid leave which is at the rate of 3½ days for each month of service. The rest of the time we officially standby during the annual refit by Thornycrofts, but obviously we won't all be needed all the time, so there should be ample opportunity for getting home. I am on a two-year contract.

We take out all sorts of bods for the various bases: geologists, wireless operators, diesel mechanics and so on. In actual fact of course they are passengers, but naturally it is easier to sign them on as supernumeraries. They have to look after their own cabins and their feeding is on a self-service basis.

We have a brand new Bell & Howell 16 mm sound projector and a library of 30 full length films, the latter being supplied by the Royal Naval Film Corporation. The electrician and I will be running that part of the business. He was also a wireless mechanic in the RAF during the war and has since been electrician on Caltex tankers. We shall be able to exchange films with RN ships out there.

The *John Biscoe* is being withdrawn next year and will be replaced by a new diesel-electric ship of the same name, and of some 1,600 tons gross. It is being specially built for the job and our captain who has been some 5 years on the present *Biscoe* is taking over the new one next year. So there will be two new ships on the job. In due course our Discharge Books will be stamped 'Antarctic Survey'. It is only a small ship, but there seems to be a colossal 'presence' attached to it. The thing I like about it is that it is not a commercial effort and hence the whole atmosphere is very much nicer. And of course one can save money as there just is no way of spending it out there.

Don't get alarmed about the purser side of the business! It is nothing like AH. It consists mainly of typing for the captain and paying out cash to the crew. I made it quite plain to the Crown Agents that I was no high-pressure accountant and they assured me that that would not be expected of me.

I had a couple of days down at the I.M.R. works at Croydon not so long ago: real red carpet treatment – car to meet me at East Croydon station and to take me back again. And free eats! Apparently Blue Funnel have got some of the new I.M.R. gear.

My own typewriter is down on the ship and I am using my mother's which is some 30 years old, but doing very well considering. We also have two new Halda (Swedish) typewriters on the ship: one with standard carriage and the other with double length manifest carriage.

We are a selected met. ship as well. Anyway, keep a look out for me on HF and let it be known that it is the Royal Research Ship and that as such I shall expect prompt service – none of this [23]QRY 11 stunt! On the other hand don't give me an RN operator to deal with!

Dick added the last paragraph because I was working at Portishead Radio Station and, as this was the UK's long-range station, it was the one he would communicate with.

---

[23] *Q Code for 'your turn is'*

LETTERS 1956-58

Chelsea, 25th August, 1956

We arrived in Southampton at the end of May, and the *Shackleton* will be leaving for the South again in October, but I won't be there. The Old Man and I are going on the new R.R.S. *John Biscoe* which was launched at Paisley on June 11th: we are due to take delivery of the ship on November 19th. We shall be going round to Southampton to load stores and bods for the Antarctic. It is intended that both our ships should be at Deception Island when the Duke of Edinburgh visits the place in the Royal Yacht. We may also take the Governor down to Vahsel Bay to visit the I.G.Y. Base there at Halley Bay.

The new ship is being specially built for the job and is a bit bigger than the *Shackleton*. The G.R.T. will be around 1,200 and the overall length will be 220 feet, as against the *Shackleton's* 180. Propulsion is diesel-electric and it is hoped that 12 knots will be reached, as against 10 of the other ship. The hull plating will be much stronger which is a good thing as we knocked out some rivets from the bows of the *Shackleton*, which was not such a good thing. As the ship was designed and built for the job, they have had every opportunity of making it an ideal ship, though that would probably be too much to hope for – funny things happen when Government departments go into the shipping business! Being built in the UK, I wouldn't be surprised if the accommodation in the new ship won't be quite as good as on the *Shackleton*.

One encouraging thing is that I have had quite a say as far as the radio installation is concerned. It is going to be the new Marconi stuff: Globespan transmitter, Reliance emergency transmitter, and two Atalanta main receivers. I asked for two main receivers, and I am getting them. There will also be a guard receiver and an auto-key for SOS and [24]AA signals. There will be a broadcast receiving installation. The main transmitter is a 400 watt affair; A1, A2 on M.F.; A1, A2, and A3 on I.F.; and A1, A3 on H.F. That means we have long distance H.F. R/T. It can accommodate 50 crystals giving 70 frequencies. The transmitter can be remote controlled for R/T from the bridge.

I would have liked to have had D/F for the sake of completeness, but high policy has decided against it: the only time it would ever be used would be in the Channel. Anyway, I won't be called out in the middle of the night to take bearings.

The main receivers look rather like the Electra. You may remember that at one time Marconi's fitted the twin receivers – Mercury and Electra: they both covered M.F., but one was essentially L.F., and the other H.F. with calibrated bandspread on the marine bands. Well, the Atalanta gives complete coverage from 15 kcs. to 28 mcs., and above 800 kcs. it is a double conversion superhet. It also has the calibrated bandspread. Normally one of these sets with a fixed tuned M.F. guard receiver is considered sufficient, but for my purpose I must have two sets covering H.F., so that I can monitor two stations, i.e. Area Station and Port Stanley or HMS *Protector*, to say nothing of the Bases.

What really amuses me is dealing with Marconi's in the capacity of a customer instead of an employee. I have been to the London office to see their contracts man, and next month I shall be going to Chelmsford to get the dope on the gear. I shall probably to going to the Marconi Glasgow depot who will be fitting the stuff. Then sometime in October we shall all

---

[24] *Auto Alarm*

be gathering in Paisley ready to take the ship over and move into our new home.

Radio conditions are very uncertain in the Antarctic: sometimes one can achieve wonders, and at other times things are hopeless. I worked Portishead on 12 mcs. from south of the Antarctic Circle, in fact I worked GKG several times from Antarctica.

I was glad to hear you identify yourself on those two occasions: pity it wasn't possible to have a chat! Royal Mail Lines are our agents in Southampton, and they pay us while we are in the U.K. (*It was surprising that, as there were over 90 operators at Portishead, I should communicate with him twice.*)

The old *John Biscoe* has been taken over by the New Zealand Navy, and renamed *Endeavour*. She sailed from London a few days ago. She will be operating from New Zealand for the Commonwealth Transantarctic Expedition.

By the way, Area 1A frequencies were crazy on our way home: only one 8 mcs. channel broadcasting when 16 mcs. should have been used as per 1B which was blasting in. I had to [25]QSX 1C for several days after actually crossing into 1A. No end of ships were calling "CQ [26]QRU? 1A", and not because they had all overslept either!

---

[25] *QSX – I am listening to…*
[26] *QRU? – Have you anything for me?*

Chelsea, 21st September, 1956

This ship (*the new John Biscoe*) has been specially designed as an Antarctic ship right from scratch, and therefore they had every opportunity to design a first class job, instead of which it has turned out to be a typical British bodged up job of half-measures. That is no reflection on the Clyde shipyards, but on our prize consulting naval architects who did the designing!

The ship has pleasing lines, and is a very solid job. The first 40 feet of hull is of one inch plating, and the rest of ¾ inch. There are also lots of additional frames. She was launched in June, and I went up to Paisley to see it at the beginning of this week.

The internal layout is very badly designed with lots of wasted space, and yet the money has been put into it. For instance the accommodation bulkheads are veneered panelling, which I did not expect, and there are proper full sized baths in the bathrooms. Needless to say, no thought whatever has been given to the radio side of things. The wireless room adjoins my room in a house of its own at the after end of the boat deck, and believe it or not, but THERE IS NO INTERNAL ACCESS! That on a specially designed Antarctic ship!! Adjoining the wireless room is a FAN ROOM which was slapped on as an afterthought. The wireless room is far too small, and yet my room is bigger than need be. If the bulkhead had been moved a foot it would have made all the difference in the world. Apart from the skipper's room, I have the only decent sized wardrobe in the place, and even then there is a lot of wasted space. Due again to the crazy layout, there isn't a decent sized settee in the ship.

Again, apart from the skipper, only the mate and chief engineer have tiny typist sized desks with drawers only down one side. The rest of us have stupid little chests of drawers with a little pull-out flap to write on – they would break off under the weight of a typewriter! There again, it is not due to lack of space, but bad use of same. All those who have seen the ship are hopping mad about the inside thereof. The *Shackleton* which was a modified ship is much better thought out as far as internal arrangement go – probably because it was more or less entirely left to the Danish shipyard. I bet the new Danish Antarctic ship, the *Magga Dan* will knock spots off anything we can produce in the next 100 years.

Grosvenor Hotel, Glasgow, 10th November, 1956

We completed our trial on the Clyde, reaching a speed of 13.6 knots. The ship is now in the K.G.V. Dock in Glasgow, and is tied up ahead of the *Calchas*. We will be moving on board on the 16th, and we leave for Southampton at 9am on the 17th. Then we leave Southampton on the 26th.

I thought the ship would be complete in every respect for the trials, but that was by no means the case: joiners and painters still have lots of work to do. They are having to run another cable all the way from the engine room to my room for an electric heater! You see, the accommodation heating is by hot air (a horrible system), and I am at the end of the air line with the result that if the air is made warm enough to heat my place, everybody else is cooked alive. The blower outlets are of a design which cannot be completely shut off. And as I think I told you before, my place is at the after end of the boat deck and has no internal access. It is a good job the trials weren't in summer, or the inadequate heating of my place would not have come to light until in was too late!

There is enough space wasted in the accommodation to build another ship out of. In fact the thing wasn't designed; but it was perpetrated!

However, there are consolations: the hull seems immensely strong – a very heavy ship for its size, there is plenty of propulsion power, and I will be able to keep warn in my dog kennel.

In many ways the ship is quite luxurious, but the design of details is hopeless, and that applies to every department. For instance the sinks in the galley and pantries are for too small. The electrician and chief steward have already made out long defect lists for next year's refit. It seems the ship has cost rather more than was expected, with the result that they are now cheeseparing. But when we come back next year we could ask for twenty or thirty thousand pounds' worth of work to be done, and we would get it without as much as a murmur.

R.R.S. *John Biscoe*
Port Stanley, Falkland Islands
18th April, 1957

About the only weekend that I could have travelled from Glasgow to visit you was occupied in moving my digs – accommodation was a great problem while I was there, so unfortunately that was that. Although there wasn't really much for us to do during the final stages of the ship's fitting out we still had to be present. We had to do a fair amount of last minute chasing to make sure jobs were at least finished off, even if badly finished. Things like no handles on drawers, washbasins loose on their fixings, leaking taps, door hooks, and a host of small details. Joiners were still working in my cabin the day we actually signed on, and they didn't leave till 5 p.m. I was then up till 2 a.m. to get the wireless room into some sort of order just for the trip round to Southampton.

This must be one of the biggest white elephants ever to come out of a British shipyard. The design (not the fault of the builders) is nothing short of crazy: if it had just happened it would have been much better. As it is, all one can say is that it was perpetrated. On top of that, the workmanship is nothing short of lousy. Typical British post war "couldn't care less and we'll strike if you don't like it" sort of finish to the whole thing. The workmanship of the actual hull is not too bad, though the design has proved it inadequately strong for the job, and there has been extensive damage. The internal finish and woodwork is chronic. The people who built the thing normally build dredgers and hopper barges, with the occasional tug or floating crane thrown in, and any sort of accommodation workmanship is something quite foreign to them.

I went on board the *Shackleton* (my last ship) a few days ago, and it is a revelation to see how much better Continental workmanship is than British. The other ship is a real business-like looking effort even though she is rather underpowered – only 800 H.P., while we have 1450 - but this thing is some 40 feet longer. The *Shackleton* has a huge hatch opening which is mechanically opened by a travelling crane – a wonderful gadget. Here we have the good old fashioned derricks with wooden hatch boards, and rather a small hatch on top of all. To look at this thing there just is not one modern feature about it: it is a typical coastal passenger ship of around 1930 complete with wireless room on the after end of the boat deck with no internal access! It has turned out to be a bitter disappointment to all of us, and even more so after all the song and dance about the last word in specially designed Antarctic ships. There is probably going to be a pretty good row about it when we get home. The thing cost a round half million, and the preliminary estimates for repairs and refit are given as £80000 – yes, I mean eighty thousand pounds! It may well be more. As it is public money that is being spent, it is well worth having a row about.

They go to the trouble and expense of having all electric

engine room telegraphs, and then they stick the wireless room as far from the bridge as possible and don't even provide an inside access!

The ship was handed over by the builders in a disgusting condition. Another contrast with taking over the *Shackleton* from the yard in Denmark – it was like taking over a yacht. But this was like taking over something which had been in use for at least five years and not even well looked after during that time. The only cleaning that the accommodation got consisted of two old down-and-outs going around with a sack, and a broom, and a shovel. There was paint of every description spilled on the lino decking and on the tiles in the bathrooms. The steward had to use steel wool on the deck in the wireless room and my own cabin to try to get the paint off. The lino on the wireless room table looks like something from a third class railway carriage floor, and an old one at that. The bathtub in the officers' bathroom has been ruined by workmen walking inside it with hobnailed boots on. Table tops have been trampled over and badly scored. I could go on like this for ages. It is nothing short of heartbreaking. About the only good thing is my wardrobe which is the best in the ship, but then you need a little more than a good wardrobe in a specially designed Antarctic ship! If this ship lasts as long as five years, most of the inside woodwork will probably have fallen down as some of it is already in a shocking state.

The internal layout of cabins is pretty chronic. The mates' cabins on the *Shackleton* are smaller but better: much more has been fitted into a smaller space by careful design. But here there has been no design at all. Every cabin in the *Shackleton* has two electric plug points and an aerial and earth plug, and of course they all have desk lights – features which are lacking here. The lighting in the wireless room consisted of one deck-head light behind me – another typical British idea. I had a slight purge and they put a desk light in, but being a complete afterthought, there is no way of putting it in a really comfortable position.

My cabin is vast for the size of the ship, but it is full of wasted space which could have been devoted to the wireless room which is adjoining and which is ridiculously small. There is a large locker space under the head of my bunk which is blanked off as there is no way of getting at it because of the chest of drawers. For the same reason, there is only one drawer under the settee. The settee is too small to be of any real use. There is no proper desk in my cabin, but just a board which pulls out from the chest of drawers. Even a thing like I had on the *Glengarry* would have been infinitely better. The desk space in the wireless room is grossly inadequate. In fact, the whole thing is just crazy.

I am going to have a purge of my own when we get home about this outside door business. I am going to consult the Union (*R.O.U.*) about it in Southampton. Not only is it very inconvenient, and very uncomfortable in bad weather, but it is nothing short of plain dangerous in this part of the world. The night we left Southampton, spray was squirting into the wireless room through the door which opens straight out onto the starboard side of the boat deck. With a full gale from that side it is very difficult to move the door against the wind. But even that isn't the worst that can happen. If there is a blizzard, they have to turn out all the deck lights because of the glare in the falling snow. That means that I have to step straight out into pitch darkness as well as into a blizzard with decks and ladders covered in snow and ice. I then have to get down an open ladder to the next deck, dodge deck cargo and other obstructions, and feel my way to the door leading into the main accommodation. In the meantime, my own door freezes up! On one occasion, I had to kick it really hard to be able to get out, and then, after coming up from dinner, I couldn't open it and had to get one of the crew to get a rope round the door knob to be able to pull it open! It was very forcibly impressed on my mind coming here on the way back from South Georgia when we were taking green ones right over the funnel and my house one evening and night.

Now supposing I was knocked down the ladder by a wave like that, and put out of action, no one would even fall over me during the night as every man jack on board can get to the bridge without so much as sticking his nose outside! On the assumption that I wasn't washed overboard during the night, no one would even miss me till after breakfast the following day. Now, if that happened in a blizzard, I would be a frozen corpse by the morning. The only consolation is that I wouldn't smell in that climate! I know this all sounds highly dramatic, but it could happen so easily, and it only has to happen once! And this is a specially designed ship!

Even the new Marconi equipment has proved a bit of a disappointment. Considering it is Marconi gear, the finish is very poor and cheap, and the stuff is not as robust as it really ought to be, although of course it is all type approved. The Globespan transmitter is far too compact for the power and facilities provided, with the result that any sort of maintenance is difficult enough in port and almost completely impossible at sea. Just about at the Equator on the way out, the P.A. feed choke burnt out, and it took the electrician and myself a whole day to fix it. It was electrically a simple job, but mechanically it was a major operation. That was caused by wafer switch failure which left the 4, 6 and 12 mcs bands out of action until a couple of days ago. We got back to Stanley on the 11th, after our second southern voyage, with the main transmitter just about completely out of action. After much work on it, only the 6 mcs band is out of action and will have to remain so. Just as we left Montevideo outward bound, the second Atalanta receiver packed up completely due to component failure – good job I had insisted on two main receivers! That left me with one receiver and more or less half a transmitter (not counting the Reliance which is no use at all in these parts) for coping with the Royal Visit! Anyway, I have been working Portishead regularly all the time and seldom had to bother with Capetown. From the Antarctic, I also worked KPH (*near San Francisco*) and DAN (*Nordeich, Germany*), also ZLO (*Wellington, New Zealand*). I worked VIS (*Sydney, Australia*) from near the Cape Verde Islands.

The Royal Visit was certainly pretty hectic, but it was delightfully informal, and everybody enjoyed it. It was quite a novelty to be flying the Royal Standard at the masthead, and to be sitting opposite the [27]Duke at dinner. We met the Royal Yacht on New Year's Day south of the Antarctic Circle, and the Duke came on board around 9 a.m. He left us around 11 p.m. the following day. During that time, we took him to 4 of the Bases. Commander Parker was in the wireless room every now and again to send press reports to the *Britannia* on R/T. The Duke also came into the wireless room. The radio staff on the *Britannia* number 26, including the lieutenant commander in charge. H.M.S. *Protector* was also in company, and they are not lacking in personnel either. And here I was on my own dealing with the Yacht, the warship, Stanley, the Bases, Area Station, etc. The naval bods were astounded that this was a one man job! For those two days I was at it almost non-stop from about 6 a.m. till midnight. Even after the Royal Party's return to the *Britannia*, there was a certain amount of traffic between us for several days.

Our first Southern voyage was purely for the Royal Visit, and we were back in Stanley for the Duke's visit here. We went on board the *Britannia* in Stanley and I was shown over their 'music room' which was quite an impressive set-up. Incidentally, when we got back to Stanley it was discovered

---

[27] *the Duke of Edinburgh*

that there was between 5 and 6 feet of water in the forward hold, and most of the stores for a new base were ruined.

By the way, the Globespan has H/F R/T, and I was carrying out daily tests with the G.P.O. radiotelephone terminal in London as far as the Equator when the set blew up. That was real *Queen Mary* style!

After a few days in Stanley, we went back south for 3 months for the actual work of the season. If I never see another iceberg again, I will not be disappointed. At one base, we were so surrounded by ice that blokes were walking between the ship and the shore without any trouble. Then a large iceberg started capsizing alongside us and trying to scoop us up at the time – terrifying! When you see railings mangled as if they were putty by a passing iceberg you realise the tremendous force behind some tens or hundreds of thousands of tons of ice. We leave for home on the 29th, our route being South Georgia, Tristan da Cunha, St. Helena, and Southampton.

I have no grouse against Portishead, but Capetown is so bad I am going to put in a report to Marconi's. With the diversion of Suez traffic, they are in chaos. Our call sign is now [28]ZDLB and once I waited three hours before they got to my traffic. You can call them for anything up to two hours before getting a reply as there is only one operator on each band to deal with calls and traffic from merchant ships and warships. At Christmas, the chaos was indescribable!

The Suez Crisis began the previous October and, as the Canal was closed, ships were diverted round the Cape.

For the purposes of the Area Scheme, the world was divided into numbered areas with each one covered by a Transmitting

---

[28] *it had previously been MXDS*

and Receiving Station. When leaving the UK, Dick would notify the nearest GPO Coast Station (perhaps Niton/GNI on the Isle of Wight) as to the port his ship was sailing from, where she was bound, and that he was listening to Area 1A i.e. to the broadcasts of Portisheadradio/GBZ & GIY. Niton would pass the information on to Portishead by teleprinter and all messages to the ship would be broadcast to Area 1A until they were told otherwise. On passage to Antarctica, the Shackleton and John Biscoe crossed into Area 1C, still covered by Portishead, when they reached 25° North, but, prior to this, Dick would notify Portishead as to exact time he would begin listening to Area 1C broadcasts. South of approximately 10° South, the ship entered Area 2, covered by Cape Town/ZSJ. Dick would notify Portishead of the impending changeover and Portishead would inform Cape Town.

The broadcasts, made at specified times, began with a list of the call signs of the ships for which messages were about to be sent, and if the call sign of his ship weren't among them, the radio officer, reverted to keeping watch on 500 kcs (now khz), the calling and distress frequency. If the call sign of his ship were in the traffic list, he continued to listen until the message was transmitted. Dick's moan about waiting was justified. The traffic lists and the traffic were sent in alphabetical order and, with ZDLB as his ship's call sign, he must have had a long wait. To establish a naval presence, there were about half-a-dozen Royal Navy telegraphists at Portishead, but Cape Town was operated entirely by naval personnel.

Chelsea, 15th June, 1958

This last trip was particularly trying and particularly hair-raising, so I have now packed up. Three round voyages are quite enough, and I don't think I could face another trip. I am still on pay, and having stirred up a good schemozzle about the whole radio organization in that part of the world, the Governor of the Falkland Islands is getting his teeth into things. I submitted two comprehensive and rather powerful reports in Stanley with the result that I had to attend a conference at the Colonial Office a couple of days after we arrived back, and then I had to go back to Southampton to attend the Refit Conference on the *John Biscoe* as more structural changes will be required. One of the ideas was that I should see both ships through their refits as far as radio is concerned and possibly attend their trials after refit. As the opposite number on the other ship has also packed up, there will be chaos due to lack of continuity, and another idea was that I should hand over to the successors – if they can find them in the first place!

We left Southampton on October 21st last for St. Helena, Tristan da Cunha, South Georgia, and Port Stanley. As usual we went straight out into bad weather in the Channel. It was even rough near St. Helena, where we stopped for less than 24 hours. Then on the Tristan with cargo and passengers! The latter consisted of a schoolteacher (a woman of 50 – brave soul!), and an agricultural expert with his wife and three small boys. Tristan is a very trying place as the weather is never really good. The island is only 3 miles in diameter, and there are only two landing places, most of the coastline consisting of sheer cliffs rising to possibly two thousand feet at a stretch.

A gale started one night while we were at anchor, and we were pitching and rolling in a most alarming and uncomfortable manner. Our scow and towing boat were tied up astern of us.

They broke adrift and were driven ashore. The motor boat was a total loss, only the fuel gauge being found on the beach. The scow was very badly damaged, but was eventually towed back to the ship after tarpaulins had been wrapped round it to make it more or less waterproof, and wires and bottle screws to hold the timbers together.

Our anchors dragged, and we were dangerously near the surf when they got the engines going, and we pushed off out to sea to ride out the storm. Had we gone aground, the ship would have become a total loss. For some 36 hours we were hove-to in 55-foot high waves! Those were the official figures quoted in our met. reports. When the sea dropped to a degree when the ship could be turned safely, we turned round and made for the sheltered side of the island. Eventually it calmed down sufficiently for us to return to the settlement to complete discharging.

From Tristan we went to South Georgia – another rough passage. The weather is always vile around South Georgia. Then another rough passage to Stanley. I never thought I would be that pleased to see Stanley again! We seemed to have spent weeks battling against bad weather and never getting anywhere – which of course was true enough, as the real season's work hadn't even started.

While we were in Stanley, the *Shackleton* was involved in SOS working after hitting an ice floe which ripped the side. Miraculously the weather was good at the time, because in that particular spot the weather is usually bad. The first information about the accident came on the B.B.C. Overseas Service news. Since I first started in this outfit I have been trying to tell people that as no 500 kcs. watch is kept at Stanley (VPC) it would be possible for a ship to be sinking within sight of the Falkland Islands without being able to tell Stanley, although it could tell Capetown or Portishead. That is more or less what happened!

Due to the loss of cargo from water damage and being jettisoned, the whole base relief programme had to be replanned. This had to be replanned again later on when the *Shackleton* was out of commission again due to major mechanical trouble when she had to limp up to Montevideo for dry-docking.

After the SOS business, the *Shackleton* was escorted to South Georgia by H.M.S. *Protector*. We then went to South Georgia ourselves and became a sort of advanced operational H.Q. There was a terrific amount of traffic to and from Port Stanley: all the ordering of replacement cargo, etc. The *Shackleton* was still on the air while being repaired in one of Salvesen's floating docks. One trouble with this game is that one is never "in port" from a radio point of view except when in Stanley, although all the mates and engineers pack up as soon as the anchor is dropped.

From South Georgia we went down to the Antarctic proper, where we spent 4 solid months before getting back to Stanley for a few days, and then back again down south for a three-week quick tour of most of the bases. Then back again to South Georgia and Stanley. When we finally left Stanley for home we still had to go back to South Georgia, then Montevideo, and Southampton. Needless to say, we had foul weather going back to South Georgia, and a shocking run up to Montevideo.

Added to all the hair-raising episodes, the food this trip was nothing short of vile, and I seemed to be living on corned beef (which I hate) and chocolate! It isn't even as if one could go shore for a meal. In spite of a considerable increase in pay, I still have no intention of going back. Christmas traffic was quite hectic: I took £15 worth of outgoing traffic in one day.

One day I sent 32 [29]SLTs direct to Portishead: one batch of 20, and one of 12. That is no doubt nothing to you ex-Portishead types (*I had resigned from Portishead Radio Station in 1956*), but for a single H8 ship it is quite an effort as it is eight hours wireless watch. Don't forget that on top of all that we have VPC (*Port Stanley*) to cope with. With a Falkland Island crew and a special rate of 1d per word, they tend to write letters home by radio, and receive them! It rather annoys me when I handle traffic to and from VPC from the (*English*) Channel at 1d a word, while it still costs us 1/2d (1 *shilling and 2 pence/nearly 6p*) to GNI (*Nitonradio*) or GLD (*Landsendradio*)!!!

Radio conditions generally were very good. I used to work GKI regularly on 22 mcs. in daylight (*GKI was Portishead's call sign on that frequency*) from south of the Antarctic Circle. I even phoned up my mother on the long distance R/T from south of the Circle. Conditions were exceptionally good, and it was just like a local telephone call. That is certainly more than the [30]"Queens" ever do.

Now let's get back to the deep south again. We ourselves sustained considerable structural damage to the hull when hitting a particularly solid piece of ice: in spite of our one-inch thick plating of special steel and tremendous amount of internal strengthening.

We also had the pleasure of being caught in pack ice when it took us 16 days to get from one base to another: that same distance was covered in six (repeat six) hours later on after the

---

[29] *SLT's were Ship Letter Telegrams, forwarded by a coast station to the addressee by ordinary post, could be sent well in advance, Salvesen's whalers sent them to Portishead for Christmas delivery when they were leaving for the Antarctic in October.*

[30] *the Queen Mary and the Queen Elizabeth*

ice had cleared. You could get off the ship and walk for miles over the ice. One day we drifted 7 miles in the wrong direction with the pack ice. The area was full of rocky islands which didn't help matters. One morning there was an island dead astern of us, by lunchtime it was abeam of us, and during the afternoon it 'overtook' us. If the pack had carried us onto the island that would have been more or less the end of the ship as the prop and rudder would have been smashed up. As it is, the prop was damaged by hitting chunks of ice.

For part of the time that we were stuck in the ice we had wonderful weather: brilliant sunshine and not a breath of wind. You could go out without any extra clothing on. In fact you had to be careful to avoid getting badly burnt.

One day during this spell a full gale sprang up with driving snow. Visibility was nil, and the ship was being subjected to considerable pressure by the ice. It was pushing upwards all round the ship to a height of 4 or 5 feet. It is under those conditions that a ship can be crushed. Then one evening as the wind died down, and the murk lifted, we saw two enormous icebergs apparently sailing along on their own at quite a fair speed. They passed about half a mile astern of us. In actual fact we were being carried past them (I think) by the drifting pack ice. If we had come up against either one of those two 'Rocks of Gibraltar' that would definitely have been the end.

The last season was a very bad one for ice, and at one time it was very much touch and go whether we would be able to relieve the two southernmost bases. It was only a miraculous holding of wind and weather that enabled us to do it. In fact we only did it by the skin of our teeth. A very few days after starting north again, the temperature just shot down and miles of new sea ice formed. It was very soon thick enough for sledging over. It does not look very hopeful for next season. The lowest temperatures we actually experienced was an air temperature of 7 degrees F. and a sea temperature of about 28 degrees F.

On one occasion we were heading into an 80 knot (gusting to well over 100) gale with driving snow, spray, and visibility almost nil when going from one base to another. There were lots of enormous icebergs around as well. The temperature was way below freezing and the whole ship was caked in a layer of hard glaze ice. The wheelhouse windows were iced over so that you could hardly see through them. I couldn't help thinking about those two trawlers that were lost in the Arctic after being iced over in bad weather. In the end it got so bad that we turned round and ran before the storm to another base luckily on a reciprocal course. Compared with the trawlers, we were lucky in that we could turn without going right over, and we did have somewhere to run for. We were rolling in a most horribly alarming way for a few hours before we reached safety.

We nearly got trapped at one of the bases the approach to which is along a very narrow rocky channel – too narrow even for a small ship to turn round in. The entrance to this channel was partly blocked by an iceberg, but we did manage to squeeze by it after hitting the odd bit of rock in doing so. There is a slight bend in this channel so that you can't see the entrance from the actual anchorage near the base hut. When the time came to leave this base we chugged happily down the channel, only to be confronted with the same iceberg which was now firmly blocking the only known way out! So we had to go astern back to the anchorage while the powers that be did some thinking. The motor boat (one we inherited from the *Shackleton* to replace our lost one) was lowered and a party set out to try to find a way out over an uncharted, shallow, and rocky bit of water. They took soundings, and the ship followed slowly. At one time there was only a clearance of two feet between the ship and the rocky bottom, but we made it!

We don't see much of the whalers although we do hear the factory ships on the air if they are operating off the west coast of Grahamland. We see the catchers in South Georgia. Even when the factory ships with their attendant catchers are off the west coast of Grahamland, they tend to be much further out to sea, while we are fairly close inshore in the ice fields. I haven't decided yet what I will do next in the way of jobs. Just at the moment I feel like forgetting all about ships and Morse! I don't think I could ever really get used to the motion of a small ship in really bad weather – it has to be experienced to be believed.

The R.R.S. John Biscoe remained in service until 1991 when she was replaced by the R.R.S. James Clark Ross. She was then sold, renamed Fayza Express, and scrapped in Aliaga, Turkey, in 2004.

Chelsea, 26th July, 1958

The P & O Line went over to direct employment (*of radio officers*) while I was on the *Esso Syracuse* on the Yankee Coast run. When I came back home I did write to them, but no go. I was then 31, and they would not take anyone over 30, so that was that. In any case, I would most likely have run up against the same trouble I had when I tried to join the N.Z. Line when all was O.K. until I went for my medical. As they are in the P & O Group, I was sent to the P & O Medical Superintendent who turned me down flat on eyesight. Why, I don't know, apart from the fact that the N.Z. Line lay down a minimum standard without glasses. They don't care how well you can see with glasses, but you must be able to see something without them, and I can't see much without mine! It is all very discouraging as I don't particularly want to go back to Marconi. I think I should try I.M.R.. I even wrote to

Sir Bernard Docker three years ago, but no go there for the simple reason that the R/O on their yacht is the skipper's brother – enough said!

Union Castle also fizzled after I left Alfie: they won't take anyone from outside in any department over the age of 23. The reason they give is that promotion is by strict seniority. Then as you know, I did one trip to Calcutta with Brocklebanks, and that was enough. They are still running the 1925 coal burner I was on, and several other old wrecks. I certainly don't fancy a coast station job with the Post Office: apart from anything else it means living in digs and being shifted more or less all over the country. I should imagine that Portishead at Christmas must be enough to send one right round the bend.

Dick was again wrong about Portishead. I never knew of anyone sent to a place he didn't want to go to and usually he had to wait years to get to the station of his choice. Certainly it was busy at Christmas, but the station was so efficiently and fairly run that, while no-one got the entire day off, lots were drawn and everyone had to work only a morning, afternoon or evening shift.

I think there are various reasons for blokes going down to the Antarctic bases and staying down there. Some of them are just bums, and other have a genuine interest and object in life. You do get the odd genuine scientist amongst the gash hands. Geologists from universities can usually manage a thesis after a season or two down there which will bring them a PhD degree. Surveyors who work for the Directorate of Overseas Surveys also find it useful in their careers. Then you get the odd BSc or even D.Sc. doing ionospheric and earth magnetism work – all the really advanced theoretical stuff. Newly qualified doctors also go down to get a start in their careers, and some of them have done good work for the Medical

Research Council. Doctors of course get more pay than the odd £400 that wireless operators get; I think they are in the £600 region. The meteorological bods are amongst the least skilled of the "tradesmen", and it is amongst them that you get the odd bum and stiff. They only have to do a 2 or 3 week course in England to become a Met. Observer. All they do is read instruments and convert the readings into synoptic reports for transmission to Port Stanley. Some of them learn Morse in their spare time and help out with the met. skeds. Wireless operators are of variable standard. They do get the odd bloke with a P.M.G. ticket, but "failed P.M.G." is a sure qualification for a job! The pay is low, but on the other hand it is for all practical purposes that much pocket money which they can save. A lot of the chaps are enthusiastic photographers and spend most of their two years' wages in advance in Port Stanley on the way out buying cameras and bits and pieces. Most of the chaps are young and have only recently completed their Nation Service, so they do not find the living conditions too bad. Actually they are pretty good, though I wouldn't stay down there myself.

Sledging journeys would seem to me to be the perfect replica of hell on earth, and yet most of the blokes seem to enjoy them. It was on the B.B.C. news a few days ago that the Governor of the Falkland Islands announced that three blokes were missing on a sledging trek and must be presumed lost. Apparently the sea ice over which they were traveling was broken up by a gale. Horrible thought. That happened at the southernmost base, and we took the blokes down there on the *John Biscoe*. It was reported in the Daily Telegraph and Daily Express. The amazing thing is that all but four of the party's dogs returned to base by themselves.

Actually it is amazing how few casualties they do get in that part of the world. One of the reasons for my packing my job up is that to me it seems that far too much depends on good luck both as far as the ships and bases are concerned. Now the luck is beginning to run out. One of these days there will be a really bad accident involving heavy loss of life, and I don't want to be anywhere near when that happens. I am no "Antarcticist", but it seems asking for trouble when small parties of two or three go out on sledging journeys. The risks which are taken with dinghies fitted with outboard motors are quite staggering and I would say bordering on the criminal, but there it is: ignorance is bliss. People seem to forget that with a sea temperature of about 28 or 30, immersion of a very few minutes is fatal. Apart from anything else, once you get wet, your clothes will freeze, and that again will be fatal. So you can just imagine how much comfort ordinary lifeboats are in that part of the world. On this next coming trip they are going to carry these new self-inflating rubber rafts which are covered over and give protection from the weather.

Chelsea, 30th November, 1958

I am in the middle of packing and sorting out possessions as we are moving (*to Reigate*) on the 5th December. My mortgage has come through, and I have even got a job! That is with Clan Line who are going over to direct employment. They pay about 5% above N.M.B. rates, but that isn't everything. They are starting me as from 1st January in the 9-12 year seniority scale at £68 per month. That is some difference to Marconi's £44 to £53 according to tonnage. This was too good an opportunity to miss, and came about quite by accident.

The R.O.U. suggested I wrote to P & O, N.Z.S., and Union Castle if I was interested in direct employment. I wrote back and told them that I had no luck with those companies for one reason or another some years ago. They suggested it would be worth trying again as radio rules had changed since then. I almost didn't bother, but in the end I wrote to all three and

[31]spoke them for yobs. P & O asked me no as I am over 30; N.Z.S. asked me no because of my 'seniority'; finally Union Castle asked me no as they won't take anyone over 24, but went on to say that Clan Line, an associate company, is in the process of going over to direct employment and that if I was interested they would be pleased to let me have details. They sent me an application form at the same time asking me to telephone them to arrange an interview, which I did. I saw the Union Castle Radio Superintendent on Thursday morning, and by the first post this last Friday I had their confirmation in writing! It was all very friendly, and he as good as told me that he would do his best to get me a good ship on not too long a run.

---

[31] *See under 16th July, 1954.*

LETTERS 1959-61

Reigate, 3rd March, 1959

So far my experience of the "Scottish Navy" has been pretty hellish, but as I have only been coasting on two ships, I am not allowing myself to get discouraged yet. If and when I do get a half decent ship of my own, it will seem real luxury!

The *Clan Mackenzie* is an [32]"Empire" boat (steam), and is getting pretty ripe. At least it is now an oil burner, though built as a coal burner. Clan Line have no coal burners left thank goodness. There was no running water in the cabins with the result that most of the old washbasins have been removed. That just left the officers' bathroom which had a square bath tub and two washbasins. There was a prehistoric steam water heater which needed a chief engineer's ticket to operate. It eventually produced plenty of very hot water – clouds of steam if you weren't careful – with lots of banging noises. The water was a bit on the rusty side and full of scale from the pipes, but at least it was water.

From Dundee we went to Immingham where the last bit of cargo was discharged: ilmenite ore. It was freezing cold and the hydraulic cranes on the dock were frozen up. That didn't worry me though. I was more concerned about our wash water being all frozen up. There was no sanitary water either for a while.

We then went to Middlesbrough for dry-docking and to commence loading. I thought the dry-docking was going to be a quick washing and painting job, but we were in it for 5 days. They seemed to be doing what amounted to a major refit. The prop was taken off, the shaft drawn, and all sorts of things.

---

[32] *A type of wartime-built ship.*

Then the ship was fumigated. We stayed in a hotel overnight while that was done.

From Middlesbrough we went to Antwerp for more loading. Thank goodness we had radar and Decca navigator as it was foggy most of the time. Our departure from Antwerp was delayed some 12 hours by fog. From there we went to Hamburg. Our departure was again delayed by dense fog. When we did leave we got part of the way down the Elbe and had to anchor for nearly two days because of dense fog. The number of ships held up in the river was terrific. The *Glenearn* was in Hamburg with us. When we finally cleared the Elbe, she flew past us while we were doing at least 10 knots flat out!

We then went to London, to the Royal Albert Dock, amongst all the passenger ships. I then left the fine vessel after a month on board and came home to wait for the next job. I was told I would have plenty of notice for the next ship. As it happened I wasn't home for 24 hours before I had 24 hours notice to join the *King David* at Barry Docks for another coasting trip. Anyway, they did apologise for the shortage of notice, and said it would be only a short trip. It was – only 15 days, but more than enough. That really was an experience.

When the Clan Line bought up the Union Castle Line, they collected two Union Castle subsidiaries: Bullard King Ltd., and King Line Ltd. Bullard King only had 4 ships and was quite a respectable little concern trading to South Africa, but King Line is a pretty notorious sort of tramp outfit. They have about 11 ships, about half of which are quite new. They tramp all over the world, and that is why one does not hear much about them. They carry bulk coal or bulk grain, or anything else they can get hold of.

The *King David* had been away for 14 months, and had just come in from Australia with a bulk cargo of grain: some of it for Barry and the rest for Limerick. These ships have white crews, and not very good ones at that. The ship was built in 1941: it is a 3-cylinder Doxford diesel economy job (11 tons per day) with steam auxiliaries. Radio is I.M.R. The I.M.R. bloke was fired on arrival for persistent drunkenness – he was a man of 61! The chief steward was also fire for the same reason, only he drank up the bond without keeping any records of the stuff.

The filth and general disorder on board just had to be seen to be believed. You would never have believed that human beings could have lived there. Luckily the Clan Line catering superintendent was on board, and he kept ringing the London office about the appalling state of affairs. He said that he would never have believed it if he hadn't seen it for himself.

There was of course no running water in the cabins. There were basins in some of the cabins, but they were never used. That meant there was one washbasin in the officers' bathroom for 5 of us. That one basin was indescribably filthy, and neither the hot tap nor cold tap worked. You had to get water from the galley in a bucket! That was partially remedied after the first day or two and we had hot water and a trickle of cold. The state of the W.C. was indescribable. Apparently this state of affairs was not considered unusual by one or two 'officers' who were accustomed to that kind of ship! Heating and lighting failed periodically, to say nothing of steering and main engine!!

From Barry we went to Limerick to complete discharging, then round the north of Ireland to Glasgow where I departed from the ship at a high rate of knots. I can now say I have been on a real tramp! I am now hoping for a nice new ship as a permanent job.

Reigate, 30th July, 1959

Your letter was waiting for me when I got home about a week ago. My mother could not forward it on to me as I was all over the place coasting: some of the Continental ports were at the rate of one a day.

It will no doubt interest you to hear that Clan Line have 'dispensed with my services'! I am sorry to lose the money, but apart from that I can hardly pretend to be sorry. In fact I was beginning to wonder how long I could stand it. I don't know how they get away with running some of the ships they still run.

I can't remember when I last wrote to you, but a couple of months ago I was coasting on the *Clan Brodie* for a month, and then on the *Clan Mackendrick* for a month. The latter being my last ship. We arrived in London (S.W. India Dock) from Hamburg on Monday the 20th in the morning. In the afternoon it was casually mentioned to me by a bloke from the office that it was intended that I should join the *King Arthur* at Leith on the Thursday, and sail thereon on Saturday for only a short trip. I naturally thought it was a coasting trip, and even then felt a bit disgusted after my experience on the *King David*. The bloke then went on to say, even more casually, that the ship was going to the Continent, Casablanca, and South Africa, and was booked for a full cargo of barley or maize or something, from Durban to the U.K. I was horrified at the idea, and 'asked them no'. It then more or less became a question of go, or else. So it was 'or else'. Apart from the loss of money, it is certainly good riddance.

Of course, it might well only be a short trip, but once they get you on articles on a tramp, anything can happen. If they got a better charter for a bulk coal cargo from Durban to Australia they would no doubt take it, and then you end up by being away for a couple of years. The ship was only built in 1953, but having some idea of the sort of crews they get in the King Line, it is probably in a disgusting condition by now. In fact the carpenter on the *Clan Mackendrick* was on the *King Arthur*, and he said never again. He told me that he would never have believed that humanity could sink so low as some of the types on that ship. To be cooped up with a bunch of bums like that for even a short voyage is not my idea of existence. I would rather sweep streets. Of course, there was the usual yarn that I was the only person available for the job, but that didn't wash either as far as I was concerned. Now they have one less person available.

The bloke who was mate on the *John Biscoe* when I was there served his time with Clan Line, and he didn't have one good word for them and got out as soon as he could. The transferability to the King Line is causing a lot of discontent amongst mates and engineers.

Clan Line must still have the best part of a couple of dozen of dirty old Empire boats whose state of maintenance can only be described as disgusting. The *Clan Mackendrick* was one[33]. If it had been flying the Costa Rican flag it would have been more to the point. It was just a dirty old tramp. The lifeboats can't have been swung out for years: they were well and truly painted into the chocks, and all the davit moving parts were painted over so it was obvious that nothing had been disturbed for a long time. The state of the deck steam pipes was disgusting and highly dangerous – in fact they actually had to renew a few bits, and believe me, when Clan Line renew anything, it sure needs renewing! A Blue Funnel marine superintendent would have condemned the whole ship within five minutes of going on board. Even the new ships are sadly

---

[33] *So was the Clan Brodie.*

lacking in maintenance. Not only do they use lousy paint, but they don't even get much of that.

It was heartbreaking, and very shaming, to be in Glasgow on a dirty old Clan boat just opposite anything up to four Blue funnel ships. I know we used to curse Alfie, but there is no doubt that Alfie really had class.

The Continental ports are real eye-openers these days: lots of beautiful foreign ships – not big liners, but just average cargo ships. If there was a dirty old tramp it was more often than not British. I shudder to think what the Germans must think when a thing like the *Clan Mackendrick* arrives in Hamburg. Some of those ships can only just do 10 knots.

It is obviously cheaper for Clan Line to pay higher wages than to maintain their ships in a decent condition. One of their own mates told me himself that the ships are 10 years out of date by the time they are launched. It is pretty obvious that the average British ship nowadays (excluding big liners) is a long way behind the average Continental ship. Some of the German and Scandinavian cargo ships are really magnificent.

I went round the north end of Scotland from east to west on the *Clan Brodie* and west east on the *Clan Mackendrick*. I was in Dundee for a couple of days on the latter ship over one weekend. The weather was pretty poor and the furthest I went from the ship was the nearest telephone call box.

All Clan Line's boats are about the same age, and even if they can get away with running them for another five years, they will all have to be scrapped more or less at the same time. Now the building programme is such that they couldn't hope to replace those ships more or less at the same time, so it rather looks as if in two or three years' time they will have many fewer ships, and therefore a surplus of personnel. The result is that they probably don't care how many people leave or get the push. Cargoes are also getting very tight: the *Clan Mackendrick* was only going out with 1500 tons to South Africa. The prop was barely covered when I left it, and there wasn't very much more cargo to come. British shipping certainly seems to be going downhill rapidly.

While in dry dock at Smith's Dock in Middlesbrough on the *Clan Mackendrick* I went over to the first of the P. & O. tankers, the *Maloja*, which was fitting out. The accommodation is really magnificent, and is fully air-conditioned. Magnificent as it was, the new Shell tankers are supposed to be even better. Incidentally, Shell are in the process of preparing to go over to direct employment (*of radio officers*).

It has crossed my mind that I was foolish to give up the *John Biscoe* job, and yet I could not have sailed last October on a 7-month voyage with all the business of house buying and moving to be settled. As it is I have had to arrange for my mother to have power of attorney for dealing with the house while I was with Clan Line as so many things had to be settled on the spot. The actual purchase wasn't completed till November. I don't know what I shall do next: I certainly don't want to go back to Marconi.

One of the mates on one of the Clan boats said a very wise thing: 'There is no such thing as a good shipping company, it is just that some are better than others.' That is very true. It is a crying shame that Alfie has made such a mess of the radio business. (*He is referring to their using 1st R/Os as pursers.*)

Dick is now with the Royal Fleet Auxiliary and his previous letter has been lost.

R.F.A. *Wave Baron*
Leith, 1st March, 1960

We are not going back to [34]Iceland, thank goodness. We are going to Rosyth on Friday to load, and will probably be leaving on Saturday, for Portsmouth and Plymouth. After that it is anybody's guess as to what happens. Whatever happens, we are certain to be messed around an awful lot. We may go on exercises with the Fleet refuelling warships at sea, or we may be freighting.

I have certainly hit rock bottom in this outfit. I have come across bums at sea before, but the bums in this outfit are in a class on their own. I certainly seem to have made a cardinal error chucking up Clan Line: that King Line ship could hardly have been worse than this, and I could always look forward to the possibility of getting a decent ship eventually. That possibility doesn't even seem to exist here.

The main thing now is to save money and do some serious thinking. The trouble is, the more I think, the more depressed I get! Unless there are vast changes in this outfit, I can hardly see myself making a career of it. If you write to me to the ship c/o G.P.O. London, you will always reach me, until such time as I resign!

R.F.A. *Wave Baron*
Devonport, 24th March, 1960

I need hardly tell you that here have been numerous changes of arrangements. Instead of coming straight down to the South

---

[34] *This was the time of the first Cod War.*

Coast from Rosyth, the powers that be decided that they might as well make use of us in submarine exercises in the Clyde area. So we went north-about to the Clyde and spent nearly a week on exercises, each night coming back to anchor in Rothesay Bay.

We then went to Portland to discharge our cargo of fuel oil, and then came here to discharge diesel oil. We then loaded a full cargo of fuel oil which we took across the river to Devonport Dockyard proper, and discharged it there. This afternoon we crossed the river again to the oil jetty to load a full cargo of fuel oil for Portland. In fact we are doing jobs which they normally use their small coastal tankers to do. I suppose we are doing it because we are at a loose end, and we can carry that much more than the little fellows. We are not leaving here for Portland till Monday. I believe there is a total of 40,000 tons of oil to be shifted from here to Portland.

The whole thing is an incredible waste of time: for all I am doing, or am likely to do, I could have had the best part of a week at home. Tomorrow we are signing off and signing on: new six-monthly articles are being opened, so naturally everyone has to be here for that.

Mind you, I can't really grumble about not being at home, as it is not all that long since I left home, but it does all rather seem a waste of time. I would much rather be on a proper voyage than this sort of messing around. One seems to have all the disadvantages of being at sea, with none of the advantages.

While we were still at Rosyth, one of the freighting tankers came in, and was out 24 hours later – to Trinidad, which would have been very nice. It was the *Bayleaf* – ex *London Integrity*. The peculiar thing is that in spite of the fact that those are some of the best ships in the outfit, up to the present they carry Marconi men as normal commercial ship would. In fact, they are run as commercial ships from a radio point of view. Of course, they may gradually be getting rid of the Marconi men as they still seem to be taking on new chaps in spite of apparently not having any jobs for them. It is all very strange.

One of this outfit's few good ships is in here for a big refit. She is the *Resurgent*, a fairly new ex Butterfield & Swire passenger ship. She is a sort of store carrying ship, and looks a nice job. I believe a lot of the original passenger accommodation is being used for officers. Also I understand she has a Chinese crew, which is a good thing. I didn't bother going on board in view of the amount of work going on over there, apart from which the chief sparks over there is the bum who used to be chief sparks over here. That is the chap after whom "The Boy" (my "boss") is taking after so successfully. That bloke even seems to be looked upon as a bum by old hands in this outfit, and oh boy, that must be something! Needless to say, he only has a 2nd Class ticket. In fact it would seem that 1st Class tickets are few and far between in this concern. There is certainly no incentive to try to improve oneself, or even to do one's best. I just cannot understand the policy in high places.

One consolation is that I shall be due for leave in the summer, and I can only hope that things will take a turn for the better by then. In fact I might even go and see our gaffer at the Admiralty while I am on leave and have a heart to heart chat with him. I will just have to put up with this until I am on a bit more of an even keel financially, but these conditions for a life's career – no can do.

I am even beginning to contemplate thinking about the possibility of investigating whether there is going to be anything doing on the *John Biscoe* this coming season. So you see, I am just about at the end of my tether.

Once again, looking on the brighter side of things: these ships from a radio point of view are for all practical purposes warships, and are equipped as such. So that is something one would not get on a commercial ship. Even the "Queens" would never be fitted with radio-teleprinters. An experience like this is never wasted provided of course that it doesn't last too long!

From what they say on the B.B.C. news about this law of the sea conference, it looks very much as if we are going to agree near enough with Icelandic demands. That will mean no more Iceland patrols – a good thing. One session was quite enough for me. Some people have been on it for over a year. Of course it is a good thing for anyone living anywhere near Rosyth or Edinburgh, but for anyone south of the border it is rather a dead loss. You have no idea how I appreciated that weekend at your place: it gave me strength to face this old wreck again.

R.F.A. *Wave Baron*
Malta, 7th August, 1960

Just a scribble to let you know that we are due to leave here on the 9th, and should be arriving in Plymouth on the 15th.

Today I put in my official application for relief on arrival in U.K. I had hoped to put up with things till November when the ship goes for refit, and thereby get Christmas at home, but the forthcoming session after this ship gets home is such that I have decided to make a strategic retreat.

Malta is stinking (literally) hot at this time of the year, and I shall be very glad to be on my way home. I wonder if I will ever succeed in making a successful break from the sea? – talk about the "Flying Dutchman".

Reigate, 7th September, 1960

The main point of this letter is because I won't be able to say very much on the telephone from the ship – that is assuming we are alongside in the first place and not moored out in the Forth. In this case, unfortunately, my No.1 is no boozer: he is a remarkably temperate gentleman for the R.F.A., and extremely conscientious and super-efficient. For that very reason he is certain to have his nose stuck into the books over the weekend, and he is pretty certain to expect me to do likewise! The trouble is I don't know when we are actually sailing. At the moment the biggest boozing is amongst the engineers and one or two of the mates. The actual navigators and the radio department are remarkably sober, and I might even add efficient!!!

The R.F.A. ships are by no means all tankers: replenishment at sea is not only fuelling, but armaments and general stores. Don't forget the Americans maintain a whole fleet in the Mediterranean without a single shore base. All replenishment is done by the 'fleet train'. The R.F.A. has store ships, armament ships, air store ships, and of course tankers. The [35]"Fort" boats are the old wartime ships, but they certainly still look very good from the outside, and I have good reports of their insides. They are of course all oil burners now, and some of them have quite a lot of passenger cabins built into the 'tween decks for carrying naval personnel and their families. The *Fort Charlotte* was in Hong Kong when I was there on the *Glenearn* in 1951. I think there are about half a dozen in use: *Fort Sandusky, Fort Duquesne, Fort Beauharnois, Fort Constantine, Fort Dunvegan, Fort Langley,*

---

[35] *A type of wartime ship built, almost exclusively, in Canada.*

Fort Rosalie, and of course *Fort Charlotte*. That seems to make a total of eight.

The "Forts" will be going out of service in the next few years. There are three relatively new ships now in service: *Resurgent*, *Retainer*, and *Reliant*. The first two are fairly new Butterfield & Swire ships taken over and converted. The last one is ex-Ropner's *Somersby*, a cargo-passenger ship. I think it must have proved too good, too fast, and too expensive for Ropners, and hence they sold it – no doubt to great profit to themselves. I can't remember the original names of the other two.

These new ships have had numerous alterations made, such as lifts in the holds, etc. They also have a separate naval wireless room as well as the Marconi room, and are equipped with radio-teletype. The "Forts" just have the Marconi Worldspan outfit plus one or two extra small naval sets: they don't do anything like as much as the "R" ships – they are only 10 knot efforts in any case. The "R" ships are supposed to be able to cruise at 15 to 16 knots.

An unusual and nice little ship is the *Amherst* which seems to do nothing apart from running between Plymouth and Malta. She is ex-Furness Withy's *Fort Amherst*[36]. She is about 3000 tons and was built as a passenger ship for running between the States, Bermuda, and I think the Caribbean. A sister ship, the *Fort Avalon* was laid up in Barry when I joined the *King David* there. Then there was real old-timer called the *Fort Townsend*. I think the *Fort Duquesne* was run by Alfie during the war.

I have been told that the R.F.A. "Forts" have been fitted with hot and cold water in the cabins. I know that nowadays in commercial outfits a Fort boat is looked upon as rubbish, but

---

[36] *nothing to do with the other "Forts"*

in the R.F.A. they are considered to be good ships. Of course, what the R.F.A. considers to be good may not be what we would call good, but there it is!

R.F.A. *Wave Sovereign*
Southampton, 2nd November, 1960

Just a scribble to let you know there has been a slight change – for the better. I am not No.1, thank goodness, but more or less a sort of supernumerary. So I don't have to worry about anything. The No.1 is the chap who has been here for the last 2 years, hence there is continuity. The No.2 is the permanent No.2. We are now due to sail on Friday, by which time I hope the wind will have dropped.

As a ship, this is in some ways even crummier than the [37]*Wave Baron*, if such a thing were possible! But with the Chinese crew gradually getting themselves organised, the accommodation is reasonably clean, saloon service is good, and feed seems reasonable.

Not even the 'new' accommodation in this ship has hot water in the cabins. I am in one of these new cabins. There is no bunk light, the desk light has been swiped, and as yet there are no port curtains. The cabin is full of hot air heating which nearly roasts me when the heating is on. They don't use the steam radiators as they leak badly. Yet this ship had its annual refit last April or thereabouts. It rather looks as if as far as ships alone are concerned, one can only expect mediocre tramp ship stuff in this outfit.

I think the pay rises are beginning to lag behind the [38]N.M.B. rises. They keep 'threatening' us with imminent pay increases,

---

[37] *The Wave Sovereign and the Wave Baron were "Empire" ships built by Furness Shipbuilding Co. Ltd., Haverton Hill-on-Tees.*

[38] *National Maritime Board*

but that has been going on a long time without anything happening. I shall just have to see how things go in the next year or two. By the time I get back to Singapore, I shall have been in the outfit long enough to be able to tell the Admiralty what I really think of things as I shall still have nearly a year's contract to run, and also they would not be able to say I hadn't given things a fair trial. By that time they might be prepared to listen to me, even if they don't do anything! Of course, it will be even harder (if not impossible) for me to get another job by then!

From all this you may get the impression I am depressed, but in actual fact I feel very cheerful. I am just stating facts without letting them worry me too much! I must apologise for the awful scrawl. The heating has gone off altogether now, and my hands are a bit chilly.

According to the papers, the *John Biscoe* was supposed to have sailed from here yesterday: poor devils always get bad weather on sailing days. This place is full of big ships: *Queen Mary, United States, Pendennis Castle*, and a few others, and here I am on another scruffy old wreck of a 'Wave' boat. Anyway, the main thing is to save up the cash and then see what the future brings – if anything.

R.F.A. *Fort Charlotte*
Singapore, 11th January, 1961

We had quite a pleasant Christmas on board with lots of big eats, complete with liqueurs after Christmas dinner which we had at 1p.m. There was also plenty of booze, but I am glad to say no one overdid it, and there were no troubles of disturbances.

There were a few unfortunate incidents over the Christmas holiday period. A radio mechanic on a frigate was electrocuted the day before Christmas Eve. The 4th engineer of one of our small tankers based out here was killed in a car smash around 2 a.m. on Christmas Eve. A naval officer was killed in another car smash a few days before that.

I had been contemplating learning to drive a car while I am out here, but have given that idea up. The number of serious road accidents in Singapore is terrific, and so many of them are fatal. Driving lessons cost at least $10 a lesson (1$ = 2/4 in case you are forgetting!), and you need about 10 lessons. Then there is the cost of transport to and from the driving school. Then this being the Orient, there is no question of passing the test at your first attempt, no matter how good you are. A Singapore licence is not recognised in the U.K. except as a provisional licence. Hiring a car here costs at least $20 a day. Apart from anything else, I can't see myself owning a car at home for many years to come.

We have done a couple of local spells at sea. We supplied stores to H.M.S. *Bulwark* (Commando aircraft carrier) in the Malacca Straits when she sailed for Mombasa, and a few days ago we supplied stuff to H.M.S. *Belfast* (cruiser) just outside and to the east of Singapore Roads. The *Denbighshire* passed close to us that day: we were on an easterly course, and she was approaching Singapore on her way home.

We have just come back to the Naval Base after a short dry-docking session down in the Singapore commercial docks. The [39]S.H.B. dry-docks are right at the end beyond the power station and near the Cable & Wireless cable-ship jetty. The power station seems to have been converted to oil: all the coal tips have gone, and the building itself just has a stubby concrete chimney at each end – all the numerous stovepipe chimneys have gone.

---

[39] *Singapore Harbour Board*

There was the usual quota of Blue Funnels, amongst them the new *Dolius* which looks like a newer type of "A" boat. The *Breconshire* was anchored in Singapore Roads. The *Glengarry* and *Glenearn* are due here in the next month or two. We shall be at sea then, which is probably just as well.

We sail for Mombasa on Friday 13th, and are due there around the 30th. Over 3,000 miles at 10 knots! I must say the speed of this ship is really quite demoralising. We shall be in Mombasa for about a week, and then we get involved in exercises in the Indian Ocean with the Indian and Pakistani navies. This is an annual effort. We should be back here around March or April. Then we do a trip to Manila. I believe I did tell you in one of my previous letters that we shall be going on the spring cruise to Japan via Hong Kong both ways.

The Oriental outlook of "never mind" seems to infect Europeans out here as well. For several weeks we have been trying to get a teleprinter mechanic down to the ship: even though they give us a specific day, it means nothing. Apparently it is now some 6 months since Marconi's were asked for a new zero-sharpening condenser for the Lodestone D/F, and still nothing has happened.

As far as R.F.A. ships are concerned, the Marconi rental-maintenance business is just a racket: rental yes, maintenance perhaps with lot of luck! Anyway, it is a peaceful sort of life, and I can now say that I shall be going home this year – it sounds better that way.

R.F.A. *Fort Charlotte*
Mombasa, 4th February, 1961

We arrived here on the 30th January, in the afternoon. We met the *Bulwark* off Mombasa that morning and transferred stores to her. She then shoved off to Muscat or some equally salubrious joint. We leave here tomorrow afternoon for Cochin to supply stores to a submarine. We shall only be there for a day, which is a good thing as that is India. After that we meet up with ships of the Far East Fleet, and R.F.A. tankers, and indulge in exercises with the Indian and Pakistan navies. That could be quite hilarious 'making remarks' over the R/T. I believe we get back to Singapore in March.

Our stay here has been a bit of a bind: as we are pretending to be a warship, we are treated as such from a communications point of view. As the local naval station is a very small outfit, they do not take guard for ships in port. That means we have to keep full sea watches and clear outward traffic ourselves. I don't mind 2 hour spells of radio-teleprinter watch, but 2 hour spells of Ceylon (GZH) "steam watch" broadcast is terrible. Ceylon does not run a teletype broadcast – not at present anyway. Of course, we take everything straight down on a "W/T" typewriter. It would be just about physically impossible to take it down by hand: [40] PL would be bad enough, but code would be unreadable in my writing! A mere 50 odd groups would be nothing, but 400 plus or 800 plus becomes hard work!

I went ashore here one afternoon, and was not impressed. It was stinking hot. Like so many of these places it has an air of decay about it. Today is overcast and raining.

I still don't like the idea of spending the rest of my life at this job, but before I make any more changes I am going to make sure that the next job is there and settled: no more 5 month spells looking for non-existent jobs. I don't really mind doing nothing for 5 months, but I just can't afford it.

---

[40] *plain language*

R.F.A. *Reliant*
Singapore, 15 June, 1961

The Director of Stores himself will be meeting us in Hong Kong to "inspect" the ship. I wonder what will be achieved apart from the spending of a considerable sum of taxpayers' money in joyriding out to Hong Kong and home again? This ship was inspected by the First Lord of the Admiralty in Rosyth before she came out here this time, but apart from drinking gin I don't suppose very much was achieved.

When I first handed my letter to the captain he was slightly sarcastic about it, but said that he would forward it with his comments[41]. He is a great pal of my old No.1 who has been 26 years in this crazy outfit. I showed the letter to No.1 who approved wholeheartedly, but he has reached such a state of inertia after all these years that it is almost too much trouble to breathe leave alone go so far as making an official complaint! Later on the old boy rallied to my support and told the captain that he had seen my cabin and supported every word I said. The result was that the captain sent for me and said he wanted to see my accommodation. Just to show you how "interested" he is in that sort of thing, he didn't even seem to know exactly where I lived. He quickly changed his manner when he entered my "suite" and felt the slight trickle of warm stale air which was my "air conditioning". While we were up there the chief officer (I use their official term because we also have a first officer) arrived on the scene and they discussed the matter between them.

The captain then said he would hold my letter for a day or two so that he could get hold of the shore superintendent and see if

---

[41] *This was a letter to the Admiralty which he asked me to return to him.*

anything could be done on the spot. I agreed with him, and thanked him kindly.

The very next day the whole engineering department was galvanised into activity, including the chief engineer himself: a big purge on all the air conditioning units and blowers. Amongst other things they discovered that nearly all the inlet filters were blocked nearly solid with dirt – pretty "good" for a ship that has just come out from its annual refit. They also discovered a damper controlling my air supply was closed. Yet I had been assured that everything was working as well as it ever would or could and that the equipment just was not big enough for the job. Although I doubted that statement in its entirety, I wrote my letter on the strength of it.

Although my actual air supply has been increased, it will hardly be adequate when a whole lot of cabins which are empty will be occupied. There is still something radically wrong with the actual cooling plant concerned as the only time I get cool air is when it is cool outside. When it is cool outside, as it is this evening – raining and temperature down to 78, I shut off my blower and open my porthole. All my blower produces is warm stale air: it is 75% recirculated air and only 25% fresh air, which is really pretty horrible. The other morning around 6.30 I was woken up by vile cooking smells: I shot out of my bunk, shut off the blower, and opened my port to let some fresh air in. When the temperature is 90 outside, my cabin goes up to 88, and one hot day it reached 92. I certainly don't call that air conditioning by any stretch of the imagination. After a rainstorm one afternoon, the outside temperature dropped to 80, but my "air conditioned" cabin was 85!

I didn't have a thermometer in my cabin on the *Fort Charlotte*, but I don't suppose the temperature ever went much above 75 even on the hottest days because a blanket was necessary every night and even for an afternoon "kip". I know if the accommodation temperature started getting near 80 the 1st fridge engineer used to start worrying. Here, if the temperature in the accommodation is down to 80 we consider ourselves lucky. Of course we did have a very good fridge engineer on the *Fort Charlotte*. The bloke on this ship is a drunken Irishman who doesn't know the first thing about his job, but has a very good line of "blarney". The general situation is not helped by the chief engineer being a nasty ignorant little Geordie. He lives in a big cool suite, so he isn't interested. All the captain is interested in seems to be an adequate supply of gin, so there you are. A typical R.F.A. set-up: mediocrity, inefficiency, and incompetence from top to bottom, to say nothing of intemperance!!!

The state of this wireless room has to be seen to be believed: dust and grime everywhere, and the state of the equipment is nobody's business. It took me two solid days just to clean the Marconi gear: Oceanspan, Worldspan, etc. Not only was it filthy inside, but I had to use soap and water on the panels and controls to get the grime off. It wasn't just dust, but seemed to be the grime of years just caked on. For the last two or three days I have been cleaning the naval gear, and still haven't quite finished. There was even an initial wiring fault in some of the naval gear which I pointed out to No.1, and which was then put right by the dockyard electrical people. There were also faults on some of the Marconi gear: the latest one only turned up the other day – the B.F.O. (*Beat Frequency Oscillator*) on the Lodestone wouldn't work. I traced that to an intermittent fixed condenser, and am now awaiting a replacement. The Worldspan had its trip circuit wired wrong from the beginning, and local Marconi depot had to get a relay flown out from home! That was already underway when I arrived on board.

What beats me is how the hell the ship came to leave U.K. in such a state. It must take fully a whole year for a ship to recover from a "refit" at this rate. Henry Robb of Leith did the refit, and they don't seem to be up to much. They painted the wireless room and even dropped paint onto the teleprinter keyboards and got some of the keys stuck thereby. I suppose "demarkation" would prevent a painter from putting a piece of cloth or newspaper over a piece of equipment when he is working near it. I am rapidly coming to the conclusion that we are rapidly becoming a very mediocre sort of nation.

Of course, I don't suppose I actually HAVE to clean the equipment – nobody would say anything if I didn't, but it just so happens that if I don't do it, no one else will, and as I can't stand filthy radio equipment and as there is nothing else to do, I do the job myself. The aerial rig is quite crazy, but I don't know if and when I shall be able to do much about that.

The old No.1 literally does absolutely and completely nothing. I would never have believed that a human being could do so much nothing and remain even half sane. I suppose eventually when it does get too much trouble to breathe, that is the end! The old boy comes to life around 9.15 a.m. He doesn't go to breakfast. He keeps as far away from the wireless room as possible although he lives almost next door to it. If I am working on any of the equipment either by myself or with the Marconi man, he is even scarcer than ever. He is in the bar before lunch, and after lunch he sleeps till about 4.30 p.m. Bar again in the evening before and after dinner. Mind you, he is by no means a bottle artist. In fact he is not a bad old stick, and I get on very well with him. I presume at sea he will do a watch! He hasn't a clue what goes on, but he is still responsible for what happens. But that is his worry, and not mine. Yet the old boy was no fool in his younger days. He got a 1st Class ticket right at the beginning and was on passenger ships with [42]Siemens. He then went ashore as an installation engineer with Gaumont British or one of those, and was installing early "talkie" equipment in cinemas all over the country and getting good money. Then he became redundant, and then there was a depression, and then he joined this outfit. After 26 years of this outfit you just aren't fit for anything ever again! It is rather pathetic that after all those years he gets far worse accommodation than a junior engineer or a young Chinese chief steward.

I don't remember whether I told you about or doctor's cabin: it is over the galley and the temperature is normally in the 90s and has reached 100! And that is "air conditioned". It is only on the coolest of nights that he can sleep in his cabin: he has to use the hospital, which itself is pretty awful. Pretty grim for an old boy of 63. He told me that the B.M.A. have practically blacklisted this outfit because of poor conditions.

The flimsiness of this ship is quite staggering: I understand it is the lightest construction acceptable by Lloyds. Alfie would be horrified. Apparently the foredeck did split when Ropners had it, and doubling plates have been fitted. It is a real cheap and nasty tin can. It was a Geordie tramp, and will never be anything more. Yet it has nice lines, and looks flashy and smart.

If this is one of the best ships in the outfit, I think the time has come to get out. The chances of getting a really decent ship within the next 10 years seem pretty remote, but even then there will be the same system and the same sort of people to deal with.

---

[42] *A wireless company similar to Marconi and I.M.R.*

R.F.A. *Reliant*
Aden, 27th August, 1961

We left Singapore for Hong Kong towards the end of June, and about 2 days before we were due to arrive in H.K. in company with several warships, the Kuwait business blew up, and we were all directed to the Persian Gulf. This ship called back at the Singapore Naval Base to load more stores: they worked right through a Saturday night, and we sailed on a Sunday.

We had nearly 6 weeks in the Persian Gulf based on Bahrain. It was absolute hell. A ship which is alleged to be fully air-conditioned and whose air-conditioning isn't working properly is far worse than a ship which doesn't pretend to be air-conditioned. My air-conditioning was out of action most of the time: it was out of action in Singapore, and the fault was obvious, but nothing was done till 2 days before leaving the Gulf when shore assistance was obtained, and yet the ship was nearly 3 weeks alongside in Singapore.

My cabin reached 102° and yet it was only 90° outside. The result was I didn't even get proper ventilation. At sea, unless there is a strong cross wind, I get stunk out by funnel fumes coming into my port. The extreme discomfort we suffered as a result of generally incompetence was nobody's business. The saloon is hellish in hot weather. In fact the whole ship is hellish at all times. I have been keeping a record of temperatures, and I am really going to let fly when I get home, and I don't care how many toes I tread on in the process – there will be a few.

Apart from replenishments at sea, we did two tips to Kuwait itself: the first time we went alongside the commercial wharf and the second time we anchored off. The heat was terrific, but at least it was dry. Bahrain is very humid: it is like living in a steam bath and you are permanently streaming with sweat. Horrible. Kuwait looks a thriving modern town, and the new and very modern harbour installations can deal with a large amount of shipping.

As soon as we got out of the Persian Gulf all the air-conditioning was shut off – I hope for some much needed maintenance, though I doubt it. Even though it is 90° in my cabin at the moment, it is heaven after Bahrain.

I have put in my application for relief, and it has been officially acknowledged. I gave them 3 months' notice which I hope will be enough for them to find another victim. I have also told them I want to travel by sea and not by air. So all being well, I should be relieved in October or November at the latest. I specifically asked for early relief in view of the 'intolerable conditions on board'. That might have some effect.

In spite of some back-dated pay increases which have recently come through, this ship has pretty well finished me as far as the RFA is concerned. I have every intention of getting out at the earliest opportunity. Even good pay doesn't make up for lousy ships, lousy feeding, lousy runs, to say nothing of the bums you get on board!

An ex-RAF colleague of mine is a senior technical author with Ferranti in Edinburgh. I had a letter from him out of the blue a while ago asking if I would be interested in something of that sort. I then had a very nice letter from the company suggesting I go and see them when I get home for a preliminary interview without any obligation on either side. Unfortunately, I think it is all rather beyond me: to write a technical manual on a complete radar system with detailed explanations of circuits in going some! My knowledge of radar is limited to making an intelligent noise and then getting hold of the shore people! Mind you, even that is more than my No.1 can do, or will do, but that doesn't solve the problem. Then of course there is the geographical problem of working in Edinburgh.

R.F.A. *Reliant*
Singapore, 8th October, 1961

My relief arrives by air on the 12th, and we sail for Hong Kong on the 11th, so they are flying him on to H.K. and I will be signed off there. At my request I am going home by sea.

As this ship is involved in exercises on the way to H.K., they cannot sail short handed from here. If my relief could arrive 24 hours sooner I could have left the ship here and possibly gone home on the *Glenartney* which sails on the 16th and arrives in London on Nov. 16th. Originally my relief was due here on the 15th, but they managed to put him on another flight. So they have at least tried. I have no cause for complaint as I am still getting off this ship well within the time limit. The fact that I can't get off this crate too soon for my liking doesn't really arise.

There has of course been a complete lack of liaison between London and Singapore. When I sent in my application for relief in August I asked for a sea passage so as not to spring it on them at the last minute, and I naturally assumed that London would at least have written to Singapore, but no, that would have been far too clever for the Admiralty. As it was, the first thing the local people knew was a signal from Admiralty a couple of days ago saying that I was being relieved. A bloke then came down from the office and asked me if I was prepared to fly, and I said no. They prefer us to fly as it is cheaper (air trooping charter flights) and they don't lose our services for a whole month while we travel and are being paid. However, they can't compel us to fly.

Apparently the air trooping flights are pretty grim: anything up to 110 people including women and children crammed into one aircraft. Some people even get the feeling that some of the 'planes aren't even well maintained. Then you have to leave most of your luggage behind to be sent on by sea and which may take 3 months to catch one up.

It almost seems too good to be true that I shall soon be off this horrible ship. I am thinking of going ashore "for good" again – I can't stand this 3rd rate concern. I have had a tentative offer of a job at Ferranti in Edinburgh as a technical author, but I fear it may be beyond me, and I don't want to give up my "estate" in Surrey.

Dick took the job with Ferranti, but felt he wasn't pulling his weight. This, however, was not the opinion of his ex-RAF boss, but he resigned after a few months and returned to sea with the Marconi Company.

LETTERS 1962-68

Reigate, 17th December, 1962

My new start with Marconi has not proved very auspicious so far – no, I have not 'resignated' yet, but was taken ill in Middlesbrough when I joined a dirty tramp there, and was sent home by the Shipping Federation doctor. At least I got off that ship and I shall be getting Christmas at home, but of course it is a financial loss because as far as I know I do not get any pay from Marconi. After 6 months they pay you full pay, less what you get from National Insurance.

I think I did tell you in my last letter that I was going to East Ham on 29th November on which day I was officially going on pay. The rush-hour travelling was hellish: I had to get a bus before 7.30 a.m., and then stood all the way to London in a first class coach corridor – it was jammed solid. The Undergrounds in London had to be seen to be believed: I couldn't even get near the first two trains that came along, but just managed to squeeze onto the third. Eventually I got to East Ham around 9.30 a.m. I felt anything but bright: I must have been sickening for the gastro-enteritis that laid me low and got me sent home.

I sat at the East Ham depot for the whole day while innumerable forms and documents were dealt with, and my rate of pay calculated. All this had to be done in between with dealing with the comings and goings of other bods. They gave me some of my 'grade I' seniority and finally worked out my rate of pay to be £62-10-0 per month. I was expecting it to drop back to around £50 per month, so that was a pleasant surprise. Not quite my £72-15-0 with the Admiralty, which next year would be getting on to the £80 mark, but still. Beggars certainly can't be choosers. The unfortunate thing is that I always seem to be the beggar instead of the chooser!

When I was hoofed out of Clan Line and joined the R.F.A., my pay dropped from £68 to £50 exactly per month. Anyway, there is no use in crying over the past, so will get back to the present story.

By about 5 p.m. on the 29th, they got me all sorted out and told me that I could go home and that they would 'phone me when they needed me: at the moment they had nothing for me, and it certainly looked as if I would get the weekend at home undisturbed. No sooner had I got out of the building when there was a frantic shout from one of the windows and one of the clerks came dashing out after me. The Newcastle depot was on the 'phone and a panic was brewing and they wanted someone pronto. That someone turned out to be me. Newcastle even wanted me to travel up that same night, which was quite impossible as I had to go home to pack my gear. In any case the night train arrives in Newcastle around 2 a.m., and what would I be expected to do until 9 a.m.? It was decided that I should catch the 10 a.m. train from King's Cross next morning which arrives just after 2 p.m. The ship in question was the *Castledore*, built in 1956. Sounded reasonable at the time.

The next problem was how to get to King's Cross in time. There isn't a chance in hell of getting onto a an early morning train with any luggage in this part of the world, nor is there much chance of getting hold of one of the very few porters. So it meant a taxi all the way to King's Cross from home! Then, in spite of allowing plenty of time for the drive, there were so many new traffic diversions in London that I just missed the train. Luckily, there was another one at 11 a.m., but it didn't arrived in Newcastle till after 4 p.m. As things turned out, that didn't matter. Naturally I thought the ship was in Newcastle, so got a taxi and took my gear down to the Marconi depot with the idea of going out to the ship. But no, the ship was in Middlesbrough! It is just that all the staffing in that part of the world is dealt with by Newcastle: Middlesbrough only deals with equipment on board ships in that port, but does not deal with personnel.

Luckily my predecessor was at the Newcastle depot, and he helped me with my gear on the way to Middlesbrough and to the ship. The ironic thing was that he did not want to leave the ship, and I was not exactly keen on joining it. It was just that the captain got it into his thick Welsh head that the other chap did not have enough experience. True is some ways. He did seem a mild sort of oaf, but quite adequate for the job.

From Newcastle it was 1¼ hours by train to Middlesbrough, and Middlesbrough is a hellish place to join a ship. The taxi has to drive all over the railway lines in the docks and you might easily get muddled up with shunting operations. Finally you have to carry your gear in considerable discomfort and hazard squeezing between railway wagons and the edge of the quay to the gangway. The place is really a disgrace, and at long last they do seem to be doing a certain amount of reconstruction.

It was getting on towards 9 p.m. when we got to the ship. It was in a state of disorganisation with little or no catering staff on board as it had been laid up in the Manchester Ship Canal for 4 months with just a few mates and engineers on board who then had to cater for themselves. It is a tramp outfit which only has about 4 ships now, and this one was on B.I. charter. Needless to say there was no accommodation for me on board as my predecessor was not leaving till the following day – Saturday. I eventually ended up in the 3rd engineer's cabin which was vacant as there was no 3rd engineer at the time. You should have seen the filth of the place – it was really quite incredible. You would not believe that a human being ever lived there! There was no clean linen or anything, and the washbasin was caked in grease and grime. I was feeling pretty dead by then, so I just threw the sheets off the bunk, wrapped myself up in a thick dressing gown I had brought with me, and then rolled up in the blankets. Anyway, I did sleep.

Next day I felt pretty grim, but decided to get settled into my own cabin, and that took nearly the whole day. The other chap had only been on board for 3 days and had no chance of settling into the place. The filth was one again unbelievable, and all the drawers and lockers were full of rubbish. I even went ashore and bought cleaning powder and cloths from Woolworths. I lined all the drawers with new white drawer lining paper, and unpacked all my gear. The accommodation as such was really quite good, and my cabin was far superior to my cabin on the R.F.A. *Reliant*. It was well laid out and well fitted out, but filthy. Again, the wireless room was nice and spacious, well laid out, and very well equipped, especially bearing in mind that it was a tramp. But once again it was filthy, and not only that, but no publications were up to date, and it was in s state of complete administrative chaos.

I intended to get the wireless room sorted out on the Sunday, but I wasn't up to it, and slept on Sunday afternoon instead, and then went to bed by 9 p.m. feeling even worse. I was due to sign on on Monday morning, and the ship was sailing on Monday at 6 p.m. for Antwerp. Then it would be going to London, Rotterdam, Malta, Port Said, Canal, Massawa, Aden, Mormugao (Goa) and finally Colombo, where the charter ends. They would then hope to get another charter, so anything could happen after that – possibly something 'choice' like coal from Durban to Australia of something equally tedious.

All this time on board I could hardly face eating anything, and by the time I got back home I had hardly eaten anything for about 3 days. On the Monday I felt terrible, so I saw the captain and he suggested I telephoned the Newcastle depot, which I did, and they were not exactly thrilled with the idea. It was arranged that I should see the Shipping Federation doctor in Middlesbrough, which I did that morning. He took one look at me and said I was not fit to sail, and should get home and get to bed as soon as possible. He also gave me a prescription for some medicine to get right away. Luckily the Marconi inspector had come down to the ship – he was a very nice chap and drove me round to the various places in his car, and finally back to the ship. I was nearly noon when I got back to the ship, and I had to pack all my gear again!

Needless to say there are no longer any through trains from Middlesbrough to London: you have to change at Darlington. I caught the train by the skin of my teeth, and there was not much time to spare at Darlington to make the connection, especially as the connecting train was the Queen of Scots Pullman, and I had to get a Pullman ticket for 7/6! (*37½ p*) Of course, I had to pay my own fare home. It was after 8 p.m. when I got to London. The whole countryside was covered in freezing fog, and it was pretty thick in London, but not thick enough to disrupt things too much. Barely 24 hours later everything was pretty well at a standstill because of the fog and smog, so I just made it in time. It was about 10 p.m. when I got home. I then had about a week in bed, and got up for the first time a week ago yesterday, but am still not feeling myself yet.

Incidentally, the captain was a filthy old devil: he spent all day spitting all over the deck in his very nice accommodation! It is something I would never have believed if I had not seen it with my own eyes. Luckily I wasn't there long enough to find out whether he behaved like that in the wireless room, because there would have been an almighty row if he did! He looked a boozy old devil as well. Even the Jamaican steward was so disgusted that he asked me whether he should refuse to clean the captain's accommodation or whether he should see his Union first. I told him it was something I had never come across before, and that it might be better if he saw his Union first.

Once again I feel like chucking up the whole beastly business, but unfortunately I can't. Of course, now I come under the Newcastle depot, and I presume that I will have to pay my own travelling expenses to go back there. I have written to the depot manager up there asking to be transferred back to East Ham, or failing that to Southampton. In actual fact it is easier in many ways to go to Southampton with luggage than to the London Docks.

Reigate, Surrey
25th February, 1963

Marconi telephoned this morning: I am travelling to Amsterdam on tomorrow night's Harwich-Hook boat to join a Houlder Brothers tanker called the *Abadesa*. It is less than a year old and should be O.K. Next voyage is to the Persian Gulf, but I don't know whether we come back to the U.K. or Continent, or carry on further east. I have no idea what arrangement exist for mail to or from the ship, but hope to write you a proper letter eventually from the ship.

The *Hereford Beacon* which I coasted recently was a nice little ship belonging to Medomsley (*Steam Shipping*) Co. It was just over 5,000 tons and only 6 or 7 years old. That was the 'worst' and 'oldest' of the 8 ships in the company, so the others must be pretty good. It was fully air-conditioned for the tropics as well. All their ships are built in Rotterdam. There is Dutch capital in the company, and for what is really a tramp outfit, it is pretty good.

My new tanker is obviously on charter to one of the big oil companies, but I don't know which.

P.S. Ceylon is no longer an Area Station. A big new station on Mauritius has taken over the job.

Reigate, 12th March, 1963

Marconi's rang me up this morning to let me know that they had not forgotten about me, and that they had heard from Houlders about the progress of the *Abadesa*. There has been some delay, and the situation now is that provisionally I shall be travelling to Antwerp next Tuesday night, the 19th. That means that I arrive the next morning. I only hope that the ship doesn't sail the same day so that I can have a couple of days to settle in and get sorted out generally. They will be ringing again either towards the end of the week or next Monday to confirm the final details of departure.

I must say the East Ham Depot is being as decent as the system allows them to be. In the past they just went out of their way to be nasty in spite of the terrific shortage of R/Os. Of course if Houlders were a direct-employ outfit I would be on full pay the whole time, as I would have been had I still been with Admiralty. But then the Admiralty are just giving away public money, whereas private companies do have to earn their keep.

I have lost so much money during this last year that a little more will not make very much difference! I have just about got beyond the stage of worrying about such 'trifles'. Ferranti certainly proved an expensive experiment, and yet if I didn't try it I would no doubt have cursed myself for the rest of my days for missing what might have been a golden opportunity.

I must say I didn't do any sightseeing during my coastal trip on the *Hereford Beacon*. With the temperature in the 20's I hardly put my nose out of the accommodation! It was bitterly cold the whole time. Apart from that, it is a bit dull sightseeing on one's own. I'm afraid my fellow "officers" had rather different ideas of "entertainment". They would go ashore between 10 p.m. and midnight, and perhaps return in time for breakfast. They weren't averse to bringing females back on board for the night either! The day was for the purpose of sleeping off the previous night's excesses. Most of them seemed to make a point of getting pretty plastered every night. What beats me is how they can afford it: champagne at £3 a bottle and cups of coffee at 5/- (*25p*) to 8/- (*40p*) each! Of course, they all maintained they had a "smashing time and couldn't remember a thing". That's tramp ships for you, even a good tramp. Anyway, there was no fighting on board, and when one met them in the saloon they were quite civilised.

The 20000-ton tanker, Abadesa (GHVC), proved to be Dick's last ship before he came ashore for good. She sailed to the Persian Gulf, Australia and Japan and, when the air-conditioning broke down, the Captain called on him to explain how it worked to the Chief Engineer!

M/T *Abadesa*
Persian Gulf, 10th August, 1963

At long last we are going home: it seems too good to be true. We are due at Mina-al-Ahmadi (*Kuwait*) tomorrow evening and are loading for Teesport, near Middlesbrough, where we are due at the beginning of September – the 2nd, or thereabouts. Needless to say I shall be off this thing at a high rate of knots.

The crazy thing is that I shall then come under the Newcastle Depot, so that when I get home I shall apply for transfer back to East Ham. That is of course on the assumption that I don't pack in altogether! But then there is the perpetual problem of "what next?" I don't know what the employment situation is these days in the road sweeping business.

Thank goodness our air conditioning continues to work reasonably well, and although it is about 84° in the 'music room', it feels really cool to what it is outside where it is really stinking hot. Without air conditioning this ship would be worse than the *Reliant*. Once the outside temperature approaches 70° it becomes uncomfortably hot on board without air conditioning.

I am back in the spare cabin I used during the beginning of the voyage. When the heating was put on in the cold weather in Melbourne, my official cabin became uninhabitable, as was to be expected. The 'music room' reached 99° after being locked up for the night!

We have been lucky with the weather. In Australia there were gales ahead of us, and gales astern of us, but never too bad where we were at any given time. The was a terrific gale before we left [43]Kwinana, and another one after we arrived in Melbourne when the *Hobart Star* which was at the same jetty broke its moorings and was blown aground 100 yards from the seafront road. Two of our mooring wires were broken. It took 3 or 4 days to refloat the *Hobart Star*. The weather was quite good when we left Melbourne for Newcastle, and all we had to contend with was a swell, which wasn't too bad.

Melbourne struck me as a very pleasant city – perhaps because the weather was so much like home at the time – wind and rain! I preferred it to Perth which is a pleasant enough place, but for some reason it rather reminds me of Singapore. The older buildings have a sort of "tropical colonial" air about them, whereas Melbourne seems really European. It is funny to hear the radio news bulletins giving information on snow and skiing conditions in the mountains. They are getting quite sharp frosts inland. It seems Australia also had a very severe

---

[43] *Western Australia*

winter which has played havoc with agriculture in some places. Apparently Sydney had the equivalent of 18 months' rain already in the first 7 months of this year. Newcastle, as its name implies, is an industrial town, but looks quite pleasant for all that. There are shipyards and steelworks and suchlike. I was surprised to see the size of ships they build there.

The Barrier Reef Pilot joined us in Newcastle and got off at Thursday Island. I must say North Queensland looks a forbidding part of the world: mountainous, tropical, and pretty barren. Having in effect circled Australia I can appreciate the immensity of the place. The differences are so terrific that one might just as well be on another planet, leave alone still being in the same country.

While we were going up the East Coast, Perth, Adelaide, Melbourne and Hobart were all putting out gale warnings, and either Adelaide of Melbourne put out a storm warning. That is of course the reason we went that way: to avoid the bad weather off the South Coast which could easily delay us several days.

A day or so before we arrived at Thursday Island we came across a fishing boat well and truly aground on a small island amongst rocks, reefs, and things. It was German, and must have been pearl fishing or something. They signalled to us by lamp: they had a crew of 4 and 10 passengers. They wanted to transfer the latter to a southbound ship and then try and refloat the boat at the next spring tide. We couldn't just stop there, so carried on a couple of miles to where it was safe to turn round, and came back again to see what help, if any, they wanted. We were prepared to try and take the people off if necessary. Luckily they were in no danger, and that was not necessary. It would have been risky for our lifeboat crew as the sea was a bit choppy, and as all our boats are aft, we would not have been able to provide shelter for the boat with the hull of the ship. The boat might easily have been smashed against the ship's side. I got on to the radio station on Thursday Island, and also to the Government lighthouse maintenance ship which was working in the area, and also sent an official message to the authorities in Brisbane. We heard on the radio that a southbound ship had taken off all those who wanted to be taken off, a couple of days later. So all ended well. At this time of the year there is plenty of shipping in that part of the world.

Reigate, 4th October, 1963

The *Abadesa* lived up to its reputation right to the end. We had magnificent weather in the Med and, as it was delightful 'outside', they decided to shut off the air conditioning. Within two hours the temperature in the wireless room went from 72 to 92, and the rest of the accommodation in proportion. Then of course when we opened doors and windows we got all the funnel fumes! The Kelvin Hughes radar had been out of action for some considerable time for no other reason than a broken scanner driving belt, and they didn't bother getting sufficient spares! We had thick fog off the Portuguese coast, so I had to D/F. We had a rough spell crossing the Bay. Then we were told there was no berth available at Teesport for a day or two. We had to anchor in Tees Bay for a day or so.

Although we went alongside at the crack of dawn on the day of signing-off, I didn't finish with the ship till about 1 p.m. We had to carry all our luggage several hundred yards to the nearest place that taxis are allowed, and then I had to wait 2 hours for a taxi! There is one hand trolley on the jetty, and that was being used for ships' stores. There is absolutely no consideration for people leaving and joining ships in such out-of-the-way places. You would think that with the majority of the crew being changed, either the shipping company or the oil company would lay on some sort of transport into

Middlesbrough. I am sure no one would mind contributing a few shillings towards a bus. Good job it wasn't raining, because there was no shelter at all either for ourselves or for our gear where we were waiting for taxis.

To cap it all I had to go to Newcastle to report to the Marconi Depot in person and hand in the voyage papers! I didn't get to the depot until 5 p.m. so just handed in the papers, and had to go back next morning. They paid my fare from Middlesbrough to Newcastle, and my leave didn't begin till the day after I left Newcastle, so I suppose I have no legal grouse there, except that the whole system is a bit much. I stayed the night at the newly done up Merchant Navy Hotel which was quite good. Next morning I went to the Depot and they had my leave and pay all worked out, but they only paid me up till the end of August. I also had a bit of a row with the Depot Manager. He had photostat copies of all my letters to East Ham and Chelmsford. Soon after I joined the ship I reported somewhat unfavourably on the installation to East Ham, and that was sent on to Chelmsford, who in turn sent it on to Newcastle who did the installation! Some of the work was downright crude, and some of the gear was reconditioned. I told East Ham in one of my early letters that as this was an outright sale to the shipowner, it was not a very good advertisement for the Marconi Company. As all the letters I had from Chelmsford were of a friendly and pleasant nature, I can only assume that Chelmsford tore a strip off Newcastle! When the Depot Manager told me that I didn't know the difference between new and reconditioned gear, I told him that I wasn't a b----- fool and had not come all the way to Newcastle to be insulted. I also told him that the CR300 receiver (a standby set) was in real junk shop condition!

I then saw their chief inspector who went through the screeds of repair requisitions I sent in. The "Atalanta" main receiver, although new, was sub-standard in one or two respects. It is a large double-conversion superhet. What I mentioned as being the possible trouble was hotly denied by the chief inspector as being impossible. I was quite insistent on my argument, and he then got out a handbook, and it turned out that I was right. He obviously wasn't conversant with he details of the set! So a good time was had by all!

Marconi paid my fare back to London from Newcastle because I joined the ship via East Ham, but presumably at the end of my leave I would have had to pay my own fare back to Newcastle unless of course I could have been transferred back to East Ham. Can you imagine Newcastle giving me a decent ship after that episode???

Incidentally, wireless company employment is not allowed by law in Germany these days as they consider that it is an intrusion on a man's liberty! How right they are. So taken all round, I think now is the time to make a clean break of the whole business and try and become a reasonably civilised human being once more!

Dick had been a radio amateur since 1947 and, on swallowing the anchor, he got a job as an editorial assistant at the headquarters of the Radio Society of Great Britain (R.S.G.B.) in Little Russell Street, London.

Reigate, 17th November, 1963

The RSGB Exhibition went very well. There was some stupendous stuff on show, both home-built and commercial. I went up on the opening day in private capacity: I arrived there at 11.30 a.m. and left at 4 p.m. without even having had any lunch. Two days later I went in an official capacity arriving around noon and was there till 7.15 p.m. But this time I did have lunch on the house. I was making notes for a review of the RSGB stand and the RSGB station which was operating. I also covered the GPO stand and made mention of the Services and one or two commercial stands. Another chap concentrated on the commercial gear. That was on a Friday, my first day officially in the job. On the Saturday I was pounding my typewriter till midnight, and again on Sunday afternoon.

Taken all round the job seems to be going quite well, and my efforts seem to be acceptable to the management. Incidentally, I wrote the report on the I.E.E. (*Institute of Electrical Engineers*) Lecture Meeting and the introductory paragraph to the Exhibition report. All this will be appearing in the December issue of the RSGB Bulletin. We are now up to our necks correcting and checking proofs from the printers for the December issue. It is surprising how much editing some articles need before they can be printed. The editing also includes putting in the marks to inform the typesetters what sort of type to use and all that sort of thing. For instance underlining something means italics, and a wavy underlining means 'bold type'. It is quite surprising what is involved. The selection of photographs for printing purposes is quite a job in itself. Anyway, I have no immediate plans of resignating! Whether it all works out in the long run is another matter, but I haven't got that feeling I had with Ferranti's right from the first day that it was 'no go'.

Travelling is of course hellish, though I am just about used to it: 3½ hours travelling per day. I leave home at 7.40 a.m., having got up at 6.15, and I get home at 7 p.m. If I go to any technical meetings in London I don't get home till 10.30 p.m. or later. I have been to two meetings so far, and there are more in the future. However, there does now seem more purpose to life than just waiting to get off some awful ship. Rush hour travelling has to be seen to be believed. I never get a seat in the train to London in the mornings: There may be up to 19 people in a compartment – 12 sitting and 7 standing, and the corridors of the corridor coaches may easily have two rows of standing passengers. It is a fight for survival. But coming home in the evening I always get a seat in the train. Travelling alone cost nearly £100 a year!

Due to the travelling, Dick resigned from the RSGB after nine months.

Reigate, 25th January, 1965

I have now got myself a yob of sorts and I start work on Thursday the 28th. It is nothing very grand, but at least it is local: 15 minutes bus ride and 15 minutes walk. So my total travelling time per day will be one hour instead of 3½ to 4 hours.

I believe the last time I wrote was to tell you that my job with the M.E.L. Equipment Co. had fallen through after everything looked so promising. As there is a Government Training Centre in Croydon, which is about half way to London from here, I thought it might be a good idea to apply for a course in radio and T.V. servicing under the Vocational Training Scheme. They even teach you to drive on this course, as that is virtually essential for a yob in that line these days. I figured that one can always earn a living doing that. Local firms are always advertising for service engineers and offer anything up to £19 per week.

It must have been October or November that I put in my application: as the course doesn't begin till March that seemed more than enough time. How wrong I was! I forgot that I was dealing with a government department and that time and messing about mean nothing. I assumed that all one had to do would be to apply and obviously be interviewed to see that one is not completely dumb and that one can qualify under the scheme. Having given up seafaring of my own accord, I was in some doubt about the latter, but as it happens that part was all right.

When I handed in my application form at the local Labour Exchange they sprung it on me that I would have to take a pr-entry test consisting of a short mathematics paper (elementary) which would include such things as working out square roots. So I swotted up square roots. That all seemed a bit daft as all I really want is the practical side of working on T.V. sets seeing that we haven't even got one. But then of course they cater for people with no knowledge of radio theory at all.

Eventually I got a letter from the Labour Exchange asking me to attend for the exam as soon as possible at my convenience and telling me that the time allowed was two hours. By then I was doing a temporary job at the G.P.O. sorting office at Redhill Station working from 2 p.m. to 10 p.m., so I went along in the morning. They gave me a spare office to myself and gave me the paper. True enough, as maths go, it was elementary, but not having seen that sort of stuff for well over 20 years, it rather caught me on the hop. There were some long-winded and tedious fractions problems although they were quite straightforward. I did the square root (long method) and a problem involving Pythagoras, but what stumped me completely was a ratio problem, and simultaneous and quadratic equations. I just could not remember how to do them. Anyway, I did 7 out of the 10 questions in 1 hr. 55 mins.

After Christmas they told me that I would be called to attend a "suitability interview" in a few weeks' time and that in view of my weakness in algebra I would have to be prepared to answer some questions on the subject at that interview. The idea was that I could spend some time learning up some algebra. The very next day I got a letter telling me to go for an interview in less than a week's time at the training centre in Perivale (western industrial suburb on London) for which I got travel warrants. That turned out to be a ridiculous waste of time, and all that was achieved could have been by means of a simple letter to me. All this business is, incidentally, arranged by the Regional Office in London.

At Perivale I saw a bloke who might have been an instructor there. He said that as all the mornings were taken up with radio theory it was essential that I should have a knowledge of algebra so that I would not hold the class back. He then said that as it was such a short time since I did the written test, it would not be fair to ask me any questions as there was no time to have learnt up the subject properly. He asked me a question about Ohm's Law. I told him I had a 1st Class P.M.G. ticket, but that meant nothing to him at all – he had no idea what that covered in the way of a syllabus. The result was that he said he was satisfied with my electrical knowledge and practical experience, but that he would recommend on my documents that I re-applied in two months' time after doing some more study. He then sends everything back to London, and eventually I would be told that officially by the local Labour Exchange.

On Wednesday (20th Jan.) when I went to the Labour Exchange to "sign" they gave me the official dope as received from London. It was to the effect that I was acceptable provided that I improved my knowledge of elementary algebra, but that they would not be able to place me for at least 4 to 5 months. Can you beat that? I was so disgusted that I almost cancelled my application there and then, but then decided to hold my peace as I can always cancel it later on if I want to. Obviously I would have to get some sort of a yob in the meantime. In fact the whole thing is so vague that I decided that if I can get a half decent yob I would keep the yob and to hell with the course. Presumably they would have to send me to some other centre than Croydon unless I joined the course halfway through – it takes nearly a year. That means living in lodgings, not a very satisfactory state of affairs even if they do pay you for same.

Anyway, the bloke gave me an introduction card to go and see the staff manager at Foxboro-Yoxall Ltd., in Redhill, on the off chance that they might have something suitable. They are a big concern with American and European connections who make all kinds of oil refinery control equipment and process control equipment for chemical works. They have a big new factory in Redhill which was built since I moved here from London at the end of 1958. Some of their work involves electronics. So I went to see them the following day, and after two hours and seeing various people, I had a job. I had a long talk with the staff manager and also a senior (I think the Chief) electronic engineer. The latter is an ex-[44]ham so we did talk the same language. The thing that impressed me and surprised me was that they did their best to try to find where I might fit in to begin with rather than seeing what vacancies they had, if any. In fact while I was in the staff manager's office someone rang up, and he told them they were not recruiting for any department at present. They also have their own technical writing department.

It was eventually decided that a good place to start would be in the electronic instrument department. They design and build their own specialised electronic test gear for use in the factory itself, and this department builds and maintains this equipment. So I am officially a "fitter (electronic tool)" and my pay will be 7/1½d (*approx. 35p*) per hour, which for a forty-one hour (40 next year) week works out at around £14-12.0 per week. Travelling will only cost 2/- per day and lunch is available in the canteen for about 2/6 (*12½p*).

As factories go, it is a magnificent place: spotlessly clean lino-

---

[44] *radio amateur.*

tile flooring even where machines are in use. Being "instrumentation" generally, it is not a dirty factory in the way that an engineering works would be.

It was impressed upon me that one is not tied down to any one department forever: they employ 1,200 people, so a certain amount of inter-departmental movement is possible to suit various abilities, or lack thereof! They have a very good name as employers. The first 6 weeks is a probationary period, and pay is reviewed fairly frequently. It is better to start at the bottom and work up, rather than at the top and work down as would have been the case had I stayed with Ferranti!!

The hours are 7.45 a.m. (!!!) till 5.15 p.m. Mondays to Thursdays, and on Fridays they finish at 3.45 p.m. So on ordinary days it means that I shall be home by 6 p.m., and on Fridays by 4.30 p.m. I will have to get a bus at 7.11 a.m. which is half and hour earlier than when I used to go to London, but the whole business will be far less tiring.

If this job appears to have any reasonable prospects I think I will stick to it and abandon that course. Perhaps next time I won't give up a job before finding the next one!

I was actually investigating the question of going to Australia, and I have been in touch with the London office of Amalgamated Wireless (Australasia) Ltd. – A.W.A., for short – and with the head office in Sydney. They are recruiting R/Os from the U.K. to go out under the immigration scheme, and it seems they even pay the £10 towards the fare which you normally pay yourself. They also pay £20 travelling expenses if you go out by sea and £5 if you fly. Eventually one might get a job with the A.W.A. shore organisation. They don't only look after ships, but build broadcasting stations and communications stations. Of course, it means you have to do two years at least with the company, and if you want to come back home you would have to pay your own fare. Pay seems pretty good, and they pay overtime on Australian ships, but when I really thought about it I did not really care for the idea of spending the whole time on the Australian coast in ore carriers, colliers, or cattle boats. Leave and time off seems very generous, but that is all very well if you have a home out there. If you haven't, it means living in a succession of digs and keeping your worldly possessions down to the number of suitcases you can reasonably take onto a ship. Then there is the financial and domestic aspect. If I couldn't afford to live in Edinburgh and pay for a house down here, I could hardly afford to live in Australia and pay for a house in England! I know my mother would not want to go out to Australia.

Recently a technical authoring firm in Kingston (Surrey) got in touch with me as a result of my registration with the Professional and Executive Register. I went to see them to investigate fully. It was attractive from the pay point of view - £1,100 to begin with and up to £1,400 after one year. But it would just have been another Ferranti. I don't want to repeat the Ferranti mistake – a disastrous mistake financially as I would now be getting at least £83 a month with the R.F.A. as it was 5 years on the 11th since I joined them. On the other hand I would probably have been an alcoholic by now!

Reigate, 28th March, 1965

The yob is going well, and if nothing else, it has certainly given me back my self-confidence. I was rapidly getting to the stage where I began to feel that I wasn't capable of doing anything! The work is quite interesting and is unhurried – that is the main thing. We are only a small section (Electronic Tool Section) at present consisting of only four people, including our immediate boss. Although we are in the factory and are hourly paid we are not doing routine factory work. In fact even the people on actual assembly work have it pretty easy as far as factories go as it is not the sort of work that can be done on a conveyor belt system. We build and maintain electronic test gear for use in the factory; testing, setting up and calibrating of the electronic portions of the measuring instruments made by the firm. Transistors are involved, but most of the work deals with strange methods of measuring D.C. One or two types of instrument do involve the use of R.F. or A.F. bridge circuits.

Some of the production instruments operate on millivolts or milliamps, so we produce gadgets that produce things like zero to 2 mA accurately controllable and which can operate into a short circuit or 10,000 ohms. We have to produce millivolts accurately from zero to one volt in one millivolt steps. Some of the things involve the use of electronic galvanometers for null detection in bridge circuits. These use two-stage transistorised balanced D.C. amplifiers with the answer given on a centre zero 50 micro-amp meter.

Some of the circuits originate in the States and are modified over here and some are more or less produced locally. We are given the circuits and standard cabinets and more or less carry on from there. No two instruments are exactly alike, and it is up to us how we lay out the components. Sometimes panel layouts are predetermined, and sometimes we do that ourselves. In both cases it involves the marking out and drilling of the panels and plastic panels for engraving. A certain amount of 'tin-bashing' is also involved, and in some cases the work is more mechanical that electrical, so it all makes a change.

So far I have built two instruments and am working on a third. Although I say so myself, I am reasonably satisfied with what I have done seeing that it is some 20 years since I have done anything in this line of work. The insides of some of the test instruments are real "birds' nests" and that alone gave me confidence: I make a point of making the insides as neat as the outsides. My current project involves the use of a pen recorder as it is for testing pen drive mechanisms and involves quite a lot of accurate marking out of holes, drilling same, and then tapping some of them for screws. It has been a great comfort to see that I can turn out work just as good, if not better, than the next bloke! The beauty of it is that we are not rushed and can take pretty well as long as we like, so we can spend the time to turn out a decent job.

As the firm is not an electronic firm as such, our section is run just as another factory department. For that reason they talk about 'electronic tools'. Even an oscilloscope is given a 'tool number' and lives in the Tool Stores together with press tools and things like that. Even our boss is hourly paid in spite of the fact that he has been there three years and was with the B.B.C. research department before that.

I must say the mornings do seem very long as I get up at 5.45 and leave the house just after 7 a.m., and we start work at 7.45. Our lunch hour isn't till 1 p.m. However, I get home by 6 p.m. and sometimes by 5.45 p.m. if the buses are running well. On Fridays we finish at 3.45 and I am home by 4.30 p.m. It is far better than travelling to London, and I still feel reasonably alive by the end of the day. On top of all, travelling to London seems to be getting worse and more expensive. In spite of getting less pay than with the RSGB, I am actually better off financially, which isn't saying much, but still!!

The name Foxboro-Yoxall derives from the Foxboro Co., of Foxboro, Mass., USA, and Mr. Yoxall himself. The latter started the English company many years ago and is still going strong. He is an agent for the American firm and finally they amalgamated, or he sold out to them. Anyway, it seems to be to everyone's advantage as there is a lot of American money in the outfit.

As factories go, it is a magnificent place, though I must say it does get rather stuffy and airless right in the middle of the building where we are. It is one huge enclosure covering some four acres! There is a huge car park for the benefit of the 'toiling masses' which I am told cost £40,000 and the 'toiling masses' certainly make good use of it!

There has been an almost all-round increase in pay, but I don't qualify for that until my three months are up, so for the next few weeks I shall be losing just over 2/- (*10p*) per week because of the increase in National Insurance. My probationary period is over, so after 3 months I get a small rise in pay, and my boss said he will be putting in for proficiency pay for me, which is encouraging.

I don't know that I would care to do this job for the rest of my life, but it will certainly do for a while – even a long while. I suppose as I was originally looking upon it as a sort of temporary lousy job just to fill in time I didn't expect much of it, so was pleasantly surprised. It is better than having great expectations and then being disappointed as I have been so many times in the past.

I must say that on a lovely day like this I wouldn't mind being on a nice ship going down the Channel! But I think I can forget about all that now, though I must say that from a purely financial point of view it was a terrific mistake leaving the R.F.A. and going to Ferranti. On the other hand the Far East is not all that healthy a part of the world to be in these days.

Reigate, 30th August, 1965

The long silence is not due to my sudden departure to sea or anything of that sort, though I must say I rather wish it was! Life nowadays is so routine and uninteresting that there isn't really much to write about anyway.

About the only thing of interest is that I have had some increases in pay: the biggest one was in April which was some £2 per week. I am now getting the princely sum of 8/8¼ (*approx.. 43p*) per hour which works out at just over £17 per week. Then I do one hour's overtime per week which brings it up to a few pennies more that £18 per week. That sounds quite reasonable until they knock off £3-5-0 *(£3.25p)* per week for tax and insurance.

The reason for the one hour's overtime is that in July they went from a 41 hour week to a 40 hour week when the official finishing time on Wednesdays and Thursdays was moved half an hour earlier. At the same time they asked everyone who could manage it, to work the same hours as previously which would mean two days each of half an hour's overtime rates. So in effect I am working the same hours for a few shillings extra. What does annoy me is that the 13/- (*65p*) for the hour's overtime only means 6/- (*30p*) extra in my pay packet!!! So there is no great financial encouragement to work overtime.

I have now been long enough in this job to appreciate some of its disadvantages, and I don't suppose it will come as a surprise to you when I say that I am fed up with it and wish I were back at sea! The novelty of being cooped up all day in a factory building has worn off. The ventilation is virtually non-existent and the general level of mechanical noise gets rather trying.

The worst thing about the whole situation is that I feel as if I were in a trap with no way out. I would like to look upon this job as a sort of fill-in until something better turns up, but that seems to be rather wishful thinking. There is of course no question of my leaving this job unless I have another one to step into. I can't go on just walking out of jobs and hoping that I will be lucky and get another one in the not too distant future! I think my luck has just about run out when it comes to that sort of thing, and I really must consider myself lucky to have got this job. Despite the line of flannel that the personnel manager gives everyone, there is no future in it whatsoever: if you come in as an hourly paid employee, you will remain an hourly paid employee till you retire.

After ten years you qualify for their so-called pension scheme and become eligible for the firm's annual dinner when they give out long service badges and tell you what a good chap you are! One bloke in our section (he is a pneumatic and hydraulic instrument bloke) has been 15 years with the company and is still hourly paid. If he is away sick for more than two weeks he will actually receive 15 HOURS' pay!!! Needless to say I would get nothing. Even our immediate boss is hourly paid.

It seems that the policy of the company is to employ the absolute minimum of skilled labour, and to pay said skilled labour absolute minimum wages. They seem to rely on people living more or less locally who will not have to spend large sums of money on travelling, though there are a few who do travel considerable distances. I suppose I can't blame the company for that as my actual pay is still more than it was with R.S.G.B., and my travelling is only one shilling each way on the local bus. After all, business is business, and there is no sentiment in business. The result is that young people only use the firm as a stepping stone to something better after getting a bit of experience, and dead-beats like myself use it as a last resort!!!

I saw in the papers the other day that Alfie Holt is taking over Liner Holdings which consists of Paddy Henderson and Elder Dempster. Incidentally, Alfie had a spot of bad luck towards the end of last year when the *Pyrrhus* was badly damaged by fire in Liverpool and was out of service for some 6 months. A photograph in the paper showed the whole front half of the ship enveloped in dense smoke, and I think firemen were fighting the blaze for nearly 24 hours. Then just recently the new *Glenogle* was in collision with Italian liner *Gallileo Gallilei* off Singapore.

A Mr. Ogden of Liverpool is a Labour M.P. for one of the Liverpool constituencies and he is an ex-Alfie Holt R/O. I think he was at sea from about 1941 till soon after the war. Then there is an ex-Marconi R/O (I forget his name) who is a Conservative M.P. for a London constituency. In spite of that I can't see myself as an M.P.!!!

Needless to say I still haven't done anything in the amateur radio line though I am still full of hope. If nothing else, I do still have a little hope left, but even that is beginning to wear a bit thin! On which cheerful note I might as well wrap up.

Reigate, Surrey
1st January, 1966

We knocked off work at 1 p.m. on Christmas Eve and went back to work on Wednesday. It was a delightful break and it was nice to get a few days of really civilised existence. Although this last week was only a 3-day week, I was so exhausted by boredom that it seemed a month! To say that I am disillusioned with shore jobs would be putting it very mildly!

On the 28th of this month it will be a year since I started in my present job. Looking back, the time seems to have gone by quickly, but only because the routine is so deadening and every day is so much alike that no particular time stands out in my memory. So it makes little difference whether a month or a year has gone by – it is all near enough the same. I don't seem to achieve anything constructive during the weekends as it takes that much to recover from the week!

Although I can't see any escape route just yet, I certainly do not consider my present job to be a permanent feature even though I am on the princely rate of 8/10½ per hour which will go up a little at the end of this month. With one hour's overtime my gross pay per week works out at £18-8-4 (*£18.41½p*), and I take home £15-3-0 (*£15.15p*) after deductions. The worst feature of course is starting at the ridiculous hour of 7.45 a.m. At 8.30 I often think that if I were on a half decent ship I would be sitting down to breakfast! And yet I suppose I am lucky to have this job although I must say that the idea of ending my days as an hourly paid factory hand is not very inspiring!

Dick remained with Foxboro-Yoxall for 23 years, and until he retired at the age of 65 in 1988.

Reigate, 13th October, 1968

In due course I shall be sending you a copy of "Electronics Weekly" of a few weeks back which concentrates on marine gear. The equipment now coming into use, and which will be coming into use during the next few years will make the *Glengarry's* gear look as much of a museum piece as that made the old *Glenstrae's*. It won't be many years before all ships' R/T will be SSB (*Single Sideband*).

A few weeks ago the newspapers carried advertisements for R/Os for the R.F.A. and giving the rates of pay. I felt rather sore after seeing that. Juniors started at £800 odd a year, and someone with 3 years' experience starts at between £1,000 and £1,500 p.a. Seniors get between £2,000 and £2,500 p.a. By now I should have been in that category. Instead of which I now get the princely pay of 10/2½d (*approx. 51p*) per hour which is £20-8-4 for a 40-hour week. After tax and deductions I bring home exactly £16.

I went to the R.S.G.B. radio show in London on its last day. The Diplomatic Wireless Service had a most interesting stand. They were demonstrating their own radio-teletype system which can print perfect copy from a signal which is below background noise. On a test some time ago they operated the system over a 2,700 mile path using a transmitter power of 5 watts. The normal system would need several kilowatts.

Reigate, 27th October, 1968

The R.O.U. is now the R.E.O.U. It is the Radio & Electronic Officers' Union. I think P. & O. Line actually call some of their bods Electronic Officers. In that case they also look after such stuff as engine room data transmission equipment and control gear. I don't know how much time they get for ordinary operating or whether they carry an extra man for that job.

I don't think ships will ever dispense entirely with some kind of R/O as in aircraft. For one thing there is the time difference in different parts of the world. I can't imagine the captain of a "Blue Flue" being very pleased at being roused out of bed at 2 a.m. to take an R/T call from Liverpool telling him how much palm oil to load in Singapore! He could of course get his own back by 'phoning the Liverpool office when it is 2 a.m. there! But in any case it would all have to be written down on paper, or perhaps tape-recorded.

I think what will happen in the far distant future will be some sort of direct radio-teletype communication with selective calling to any one ship. Then it wouldn't matter at what time of day or night a message came through as it would be on paper. This would no doubt have to be on VHF or UHF via communications satellites. The gear would be so elaborate that you would need a whole staff of electronic engineers. Even when such a system does come into use, it may still be necessary to carry conventional gear for emergency use in the event of a failure in the big system. It will still be necessary to carry a teleprinter operator if nothing else.

For a long time it will still be cheaper from the ship owner's point of view to carry an R/O with conventional gear rather than all this satellite stuff. It only wants one or two communications satellites to be put out of action for a whole worldwide system to collapse in confusion.

There is always so much I would like to do, but never seem to get done. The trouble is that I am so utterly exhausted by the weekend that I tend to sleep rather late on Saturdays and Sundays and don't feel like doing much for the rest of the day. The exhaustion is not due to hard work – I wish it were! It is due to utter boredom and frustration and the almost complete lack of breathable oxygen in the dump I work in. Of course, getting up a 5.30 a.m. doesn't help much either! By the time I get home in the evening I feel quite dead with very little interest in any further existence. In fact if this is my lot for the next 20 years, then there just is no point whatever in any further existence!!

Next month our local buses, which are double deckers, are being replaced with one-man operated single deckers with more standing room than seats. It is going to be chaos and the service will no doubt be slowed down, so I shall probably have to catch an earlier bus to get to work on time. I think they call it progress!

I suppose those houses we looked at on the Wimpey estate in Kirkcaldy will be very much more expensive now. Prices are going up out of all proportion in this part of the world. There is a batch of two-bedroom maisonettes which start at £6,000!!!

LETTERS 1990-92

Reigate, 29th September, 1990

I can assure you that any lack of communication on my part has been entirely involuntary. For over 4 years now I haven't had as much as one day off. I usually get to bed between 11.30 p.m. and midnight, and am up around 7.30 a.m. The domestic chores are the same every day whether it is Monday or Christmas Day. I haven't used my record player for at least 5 years. I haven't read a book for I don't know how many years. I have got several years of Readers Digests which I haven't even looked at, so we cancelled the subscription after getting it for more years than I can remember. I also had to cancel the National Geographic Magazine as I never managed more than skimming through the pictures.

It is now 9.30 p.m. and I escaped from the kitchen for a few minutes to start this letter. My mother is sitting in a folding chair in front of the kitchen sink doing the washing up – brave effort. Then I have to put everything away and eventually get her to the loo and back to her wheelchair and eventually to her bedroom. It is usually well after 10 p.m. when we finish in the kitchen. I feel I am just about at the end of my tether, both mentally and physically. I have been 'confined to barracks' for so long that at times I feel I am losing my power of speech!

On top of all we are having a lot of unpleasantness from business premises adjoining who have been complaining in a threatening manner about our cats going on their flat roof and making a mess of it. (Break to go down to kitchen.)

(00.10 30th Sept.) Must finish this before going to bed.

They built an extension adjoining my property. It has a flat roof adjoining my coalhouse roof thus providing perfect access for cats. Then they put loose gravel on top of their flat roof in spite of warnings from us that they are providing the perfect cats' loo. That is what happened. The boss man verbally threatened to sue me for thousands of pounds damages for his roof and said if I didn't keep my cats under control he would take action to have them removed.

The result of all this on my mother was catastrophic. It just about reduced her to a state of complete collapse approaching heart failure, and in tears at times. I was also very distressed by the whole business. I have enough to worry about already without that sort of thing. To cut a long story short (what a hope!). As I am a member of the RSPCA I telephoned their headquarters in Horsham (Sussex) and spoke to their Director of Legal Services who is a practising barrister. He told me that the law is that the owner of a cat cannot be held responsible for trespass by the cat or by any damage caused by the cat. But of course that doesn't stop the so-and-so from accidentally on purpose harming a cat. Also the law says that it is his responsibility to protect his own property from cats if he wants to, but what action he may take must not cause any harm to any cat, or for that matter to any human. So we do have the law on our side, but it is all very unpleasant. When I pointed out to the bloke next door that in 30 years there had never been a mess on our own flat roof because there is no loose material on it, he just started shouting at me. My next move is to see a solicitor to send them a letter pointing out the legal aspect and that any further correspondence to be sent to the solicitor's office.

Dick's mother died at the age of 91 in 1991

Reigate, Surrey
8th September, 1992

I still have two cats: they are both 13 years old this year. I am very glad of their company and hope they last a few more years. They are certainly in excellent health at present. We had 10 cats years ago. Old age and sickness reduced the number to 6, and it remained at that for a number of years. Two were farmed out to the Cats' Protection League while my mother was here. That left us with 4. Then the oldest one, she was at least 18, developed a bad cancer of the bone in the lower jaw and she had to be put down. That left us with 3. Then my very special and mysterious Ginger Angus collapsed and died of kidney failure three weeks to the day after my mother died. So that leaves me with two. They are incredible characters and incredibly affectionate.

I haven't been terribly mobile recently as I have a varicose ulcer on the inside of my right ankle. I won't bore you with details, but briefly owing to my doctor not giving it the attention it should have had, it became infected and was oozing stinking pus. I had two courses of antibiotics and had to go to the surgery once a week for over a month for the nurse to dress it. I have to wear a shaped elastic stocking on my right leg and will have to do so for the rest of my life.

I remember seeing the *Rangitiki* (GSXW) in the London Docks. The ship had two funnels. The wireless room was in the forward funnel. Some people we knew went to New Zealand in that ship in the early 60s. I also remember seeing the old *Rimutaka* in London. She was the old P. & O. *Mongolia* when I saw her in Hong Kong in 1931 when we were leaving China. My mother, sister, and I were on the old P. & O. *Mantua* which we joined in Shanghai and left in Marseilles. That ship was a two-funnelled coal burning passenger liner built long before the First World War. The ship was due for scrapping but was sold to the Japanese and ran for several more years. There was of course no running water in the cabins and no mechanical ventilation – only bulkhead mounted fans. We had the old compactums all same AH *Machaon/Glenaffaric*. The deck-heads were just painted steel. That was First Class too!

To me, as an 8 year old it was all very impressive as it was the biggest ship I had been on. The other ships I had been on were the small ones belonging to Jardine Matheson and Butterfield & Swire on the China Coast. A few hours after leaving Hong Kong for Singapore the sea was getting quite rough and a 2nd Class passenger jumped overboard from the stern. We stopped and searched till it got dark, but no signs. A lifeboat was manned and swung out ready for lowering. We heard later that the captain was relieved at not having to launch the boat as the weather was such that there was every chance of losing the boat and crew. Apparently the poor devil who jumped was in the Hong Kong police and was invalided out with a bad heart condition and was more or less going home to die. Very sad.

In the Indian Ocean in lovely calm weather I took the wheel for three days running for about an hour each morning and steered to the magnetic compass with the captain supervising. It was quite something for an eight year old to be steering a passenger liner, and P. & O. as well!

Dick suffered from the crippling disease of osteoporosis. We continued to correspond by letter until 1998, but, after that, we communicated by telephone, as his illness prevented him from typing or writing. For several years a nurse called at his home to treat his leg ulcer and he had someone to do his cleaning and shopping. He became so ill, however, that he told me that he wished he would die.

From time to time, and at his own expense, he went into the Nuffield Centre in Redhill for some weeks of respite, but towards the beginning of 2004, he became a permanent resident, paying over £700 per week: and this caused him great anxiety as it meant selling his house which he had bequeathed to his niece, Alexandra.

During the first part of his residency, I could speak to him on the phone, but there came an evening when the person who took my call at the Centre refused to allow me to do so on the grounds that I wasn't a relative. Naturally, I wasn't pleased and, when I told this person, who spoke with an Indian accent, that I wanted to speak to the manager, he replied that he was the manager. I then asked how Dick was, but his response to that was that regulations prevented him from telling me.

When the first parts of my article 'Letters from a Radio Officer' (extracts from Dick's letters) were accepted by Nautical Magazine, I sent a copy to the Centre asking that, if Dick were too ill to read, someone could read it to him. At the same time, I asked that Alexandra, whom I knew visited him, be informed. But I heard nothing from either the Centre or the niece, and when I phoned the former on the evening of 24th February, 2005 a more kindly lady told me that Dick had died on Christmas Eve. He was 81.

MOLLAND FAMILY HISTORY

Dick's father, Charles Edwin Molland, was a Commissioner of Posts with the Chinese Postal Service and, during the First World War, with the rank of captain in the British Army, had superintended the work of the civilian Chinese Labour Brigade; a little known unit whose work included the digging of trenches and which was brought to Europe in Alfred Holt's 'coolie' ships.

On returning to China after the War, Charles Molland married Tatiana Ivanoff whose family, originally from Jitomir (Zitomir or Zhitomir) in the Ukraine, were in Vladivostok when the Bolshevik Revolution of 1917 took place. Mr Ivanoff was a General in the Customs and Excise. He had been offered the post of Minister of Finance in the Kerensky Government, but refused it for political reasons and, when the Revolution occurred, he and his family fled to Japan before settling in China.

Dick and his sister were both born in Peking (Beijing); he in 1923 and [45]Lorna Daphne in 1926. But the family lived also in Shanghai, Chefoo and Wei-hai-wei: the latter two towns on the Shantung Peninsula. In 1931, five-year-old Lorna became so seriously ill with whooping cough that Mrs Molland took both children to Switzerland. And, two years later, with the little girl fully recovered, they came to live in England. Their first home was on Hayling Island, in Hampshire, then, after a brief stay in Putney, they moved to Felpham, near Bognor Regis in Sussex, where they lived from 1937 until 1943.

---

[45] Lorna married a White Russian to become the Countess Ignatieff, but died of a brain tumour in 1978, at the age of 52. They lived in Switzerland and Count Ignatieff, who was a Hereditary Prior of the Knights of Malta, worked for Trans-World Airways as an interpreter. Their only child, Alexandra, married Peter MacMahon, and, in 2004, she was housekeeper to a wealthy Indian living in the south of England.

From then, until 1958, when they moved to Reigate in Surrey, they lived at 7 Smith Terrace, Chelsea.

Mr Molland had remained in China and, in 1939, when his family was due to return on the P & O liner *Canton* (GDDT), the Second World War broke out and the passage was cancelled. The Japanese invaded China in 1935, but when they entered the Second World War, after their strike on Pearl Harbour on 7 December, 1941, Mr Molland was placed under house arrest on Shameen Island, near Canton (Kuangchou). The Japanese took most of his possessions, including his radio and electric torch, and he remained on the Island until repatriated to the UK. This was under the Red Cross Civilian Exchange Scheme and he was taken to Lourenço Marques (Maputo) where the transfer took place. He later worked for NAAFI (Navy, Army and Air Force Institutes), separated from his wife, and settled in Oslo where Dick occasionally visited him.

Dick was in his second term, studying electrical engineering, at Faraday House, when the evacuation of Dunkirk took place. And his mother, fearing the imminent bombing of London, which damaged the Central YMCA in Tottenham Court Road where he had been living, recalled him to Felpham.

He then worked for a short time in the laboratory of a small firm called Landlines Equipment where the work was very hush-hush and where the senior employees were Dr James Robinson and Ernest Gardiner, the Chief Engineer and one-time President of the Radio Society of Great Britain. This was in the village of Bricket Wood, between Watford and St Albans in Hertfordshire, on the land of Lady Yule. Prior to the War, and equally hush-hush for a different reason, was the chartering of Lady Yule's yacht *Nahlin* by Edward VIII and Mrs Wallis Simpson; another person who had lived in China.

Dick then joined the RAF and, at Cranwell, where he was training to become a wireless mechanic, interviews were held for employment in secret work, the nature of which was not disclosed. He was one of twelve chosen and, once qualified, found himself working for the Inter-Services Research Bureau (the Special Operations Executive) in a secret laboratory in Old Welwyn, near Welwyn Garden City in Hertfordshire. When, towards the end of the War in Europe, the laboratory, which had moved to Watford, was closed, Dick, although holding only the rank of corporal, was put in charge of the SOE transmitting station at Henley-on-Thames.

When the War in Europe ended, he was sent to India in Force 136. The draft went to Liverpool to embark and the men were pleased when they saw the Dutch troopship Nieuw *Amsterdam* (PGGF) which they expected to board. But disappointment ensued when they were taken to the *Tamaroa* (GFWX); a Shaw Shavill ship where the conditions were abominable and they suffered torture in the heat of the Red Sea in July. The ship sailed on 8 July, 1945 and arrived in Bombay on the 28th. But, when the first atom bomb was dropped on Hiroshima on 6 August and the Japanese surrendered after the dropping of the second bomb, on Nagasaki on the 9th, he was sent to Burma to conclude his military service. Embarking in Rangoon, he returned to the UK on the Pacific Steam Navigation Company's *Reina del Pacifico* (GMPS).

A number of men who had been in Inter-Services Research Bureau, now worked for Philco. Dick was offered a job and took it, but, when, during the fuel crisis of 1946, the establishment closed, he decided on a seagoing career and commenced studying for a 2nd Class PMG Certificate at the London Telegraph Training College in Earls Court.

It was by pure accident that he began he seagoing career with Alfred Holt & Co. He and his father were on their way to the

Tower of London when they met a man whom his father had known in China, and it was this man who recommended him to Holts.

The author with Dick (right) on the bridge of the Glengarry,Colombo, Boxing Day, 1949.

Other seafaring books by Ian M. Malcolm

LIFE ON BOARD A WARTIME LIBERTY SHIP (print and ebook formats, published by Amberley)

Describes the author's wartime experiences as the 3rd Radio Officer of the Liberty Ships *Samite* and *Samforth* during a time when Britain's Merchant Navy ships were being sunk faster than they could be replaced.

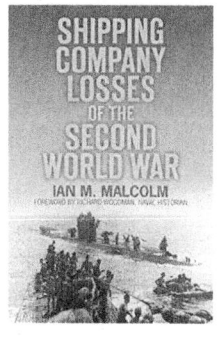

SHIPPING COMPANY LOSSES OF THE SECOND WORLD WAR (print and ebook formats, published by the History Press)

Describes the losses suffered by 53 companies in detail; giving masters' names, where bound, convoy numbers, positions when sunk, casualties and enemy involved.

DANGEROUS SEAS (Print and ebook formats due out April 2017, publisher Moira Brown)

Four book collection – *Dangerous Voyaging, Dangerous Voyaging 2, Fortunes of War* and *Mined Coasts*.

The reader will be left in no doubt of the sacrifices made by the men, and also a few women, of the wartime Merchant Navy.

OUTWARD BOUND (ebook format, published by Moira Brown. Print format due out May 2017)

The author's first post-war voyage; on the Liberty Ship *Samnesse*, managed by Blue Funnel for the Ministry of Transport. The voyage begins with calls at Piraeus and Genoa, after which months are spent tramping to various ports in East Africa and the Red Sea. A happy ship basking in the post-war euphoria.

BACK TO SEA (ebook format, published by Moira Brown, print format due out June 2017)

A voyage to the Far East on the 1911-built *Atreus*, which carries pilgrims to Jeddah on her homeward passage. The author then attends the Lifeboat School in Liverpool and stands by the 1928-built *Eurybates* in Belfast before making his first two voyages on Glen Line's *Glengarry*.

VIA SUEZ (ebook format, published by Moira Brown, print format due out July 2017)

The author makes two more voyages on the *Glengarry* before requesting a voyage to Australia prior to swallowing the anchor. He then coasts the *Glengarry*, *Elpenor/Glenfinlas*, *Helenus*, *Patroclus*, *Medon* and *Clytoneus* after which he is told that his request has been granted.

LAST VOYAGE AND BEYOND (ebook format, published by Moira Brown, print format due out Aug 2017)

The Australian part of the voyage, on *Deucalion* (built in 1920 as the *Glenogle*) proves enjoyable, but is followed by a trip round Indonesian islands, loading copra, which, although a most interesting experience, is not. On departing from the Merchant Navy the author spends three years at GPO Coast Stations.

152